APPLICATIONS OF MICROWAVES IN DAILY LIFE, EARTH, ATMOSPHERE AND SPACE

ABOUT THE AUTHOR

Dr Hari Narain Srivastava, MSc, PhD took his doctorate in Microwave Physics from Lucknow University in 1963. He joined India Meteorological Department in 1959 and rose to the post of Additional Director General (Research). He was also deputed to the Indian Air Force and later to the Ministry of Science and Technology as Director, Earth Sciences. He published more than 210 research papers in Microwave Physics, Seismology and Meteorology. He is the author of 12 books on the above subjects by reputed publishers like National Book Trust, New Age International, Raj kamal Prakashan, Ministry of Earth Sciences, Department of Science and Technology and Indian Meteorological Society. He won several national and International awards from the Ministry of Steel and Mines, Ministry of Human Resource Development, Ministry of Environment and Forests (now climate change also), Uttar Pradesh Government, International Symposium on Natural Hazards at Italy and Bulletin of Indian Society of Earthquake Technology, Roorkee. He served in several International and National Committees like International Association f Seismology and Physics of the Earth's Interior, International Lithosphere Program, World Meteorological Organisation, United Nations Framework Convention of Climate Change, Bureau of Indian Standards and many others. He was deputed to several International Seminars and meetings in Seismology, Meteorology including Climate Change.

APPLICATIONS OF MICROWAVES IN DAILY LIFE, EARTH, ATMOSPHERE AND SPACE

Dr. H. N. Srivastava
Ph.D., F. N. A. Sc.

ZORBA BOOKS

ZORBA BOOKS

Published by Zorba Books, June 2022

Website: www.zorbabooks.com
Email: info@zorbabooks.com

Title:- Applications of microwaves in daily life, earth, atmosphere and space

Printbook ISBN :- 978-93-93029-73-7

Ebook ISBN :- 978-93-93029-80-5

Zorba Books Pvt. Ltd. (opc)
Sushant Arcade,
Next to Courtyard Marriot,
Sushant Lok 1, Gurgaon – 122009, India

Printed By Thomson Press (India) Ltd. B-315, Okhla Industrial Area, Phase 1, New Delhi- 110020

CONTENTS

Dedicated to my wife, Smt Om Kumari Srivastava

PREFACE

Amongst different type of waves like sound, light, radio, X rays and gamma rays, microwaves have revolutionized our lives due to their enormous applications.The comfort of heating our food or cooking in microwave oven, communication with the cell phones at any time or place, watching television programs all over the world through dish antenna constantly remind us about their applications in our daily life. Although radars emitting microwaves were widely used since World War II to detect enemy ships or aircrafts, the advent of satellite era has demonstrated their wider applications. Microwave communication between earth and space ships has enhanced their application in space research which enables us to know more about planets, moons and other celestial objects. The support to Big Bang theory for the creation of universe has been possible through detection of microwaves in the background of sky. The use of GPS systems in the cars and other means of transport makes our life safer and easier. Infact, GPS based navigation systems have been developed for land, air and sea. Electronic warfare using microwaves in modern weapons can target enemy positions precisely. The tracking of cyclones through satellite and Doppler radar has enabled to evacuate public before they strike the coastal areas thereby minimizing loss of lives from strong winds, storm surges and floods as happened in India during very severe cyclones Phailin(2013), Hudhud (2014) and others. In fact the social and economic impact of microwaves has played a significant role in shaping our life. The book has been written in simple language so that it can be easily understood by general public. While applying laws of physics, wave propagation or other subjects like earthquakes, volcanoes or weather, due weightage has been given to the NCERT text books on Science and related subjects.

This book consists of 16 chapters. The first two chapters describe microwaves and their characteristics which are necessary to understand their applications. The third and fourth chapters discuss methods for microwave generation, transmission, reception and principles of remote sensing based on active and passive methods. Applications in daily life like microwave ovens and viewing worldwide television programs have been given in Chapter 5. Chapter 6 gives miscellaneous applications like industry, landscape, pollution and disaster management. The use of microwaves in diagnosing diseases and specialized treatments (therapy) is described in Chapter 7. Satellite and terrestrial microwave communication and its specific use in mobile phones has been discussed in Chapter 8. In this chapter, the difference in GSM and CDMA technology used in mobile phones has also been clarified. The design of GPS system in navigation and geodesy(shape of earth) is given in Chapter 9. Chapters 10 to 13 describe their application in weather forecasting, early warning systems for earthquakes, tsunami, volcanoes, landslides and glaciers. Chapter 14 gives microwave application in warfare and defence. Chapter 15 describes many applications in the exploration of planets, stars and moons revealing their mysteries. The last chapter discusses microwave hazards and mitigation.

The author is thankful to the India Meteorological Department for radar and satellite photographs of tropical cyclones and for my long association with the department for my research contributions in radio meteorology, weather forecasting and seismology and to the Department of Physics, Lucknow University, Lucknow (U.P) for commencing my research career in microwave Physics in the year 1957-59. The author expresses his gratitude to late Professor and Director Dr. Sanjiv Sinha, Head of Department, Neurosurgery, Gobind Vallabh Pant Hospital, Maulana Azad Medical college, New Delhi and Dr Shruti Srivastava, Professor, GTB Hospital, University College of Medical Sciences, Delhi for their suggestions regarding application of microwaves in medical science and to Mrs.Taru Srivastava, M.Tech.(IIT-D) Addl.G.M, NTPC, Badarpur, New Delhi about microwave communications. The author is also thankful to Sri Rajiv Sinha, Ex. Director, Ministry of Environment and Forests, New Delhi for helpful suggestions. The author would express gratitude to all the scientists and institutions whose works have been described in the book but bibliography could not be given keeping in view the popular nature of the book.

Hari Narain Srivastava

LIST OF ILLUSTRATIONS

CHAPTER I

Electromagnetic radiation and microwaves

Electromagnetic radiation is a form of energy radiated in the form of waves. These waves travel at the speed of light ($3x10^8$ meters per second). As the name suggests, they possess properties of electric as well as magnetic fields. Both these fields oscillate at right angle to each other. Electromagnetic radiation is divided into different parts (sub sets) according to wave length. The subsets are classified in order of decreasing wavelength and increasing frequency. Accordingly the nomenclature is radio waves, microwaves, infra red radiation, visible light, ultra violet radiation, X-rays and gamma rays. Table 1 gives their wave length and usage. The group of frequencies in each type of waves is called frequency band. The distance occupied by one complete cycle of such an alternating wave is called wavelength. It is equal to the velocity of light ($3x10^{10}$meters per second) divided by the frequency (number of cycles each second). The frequency is expressed as hertz (cycles), kilohertz (KC) or megahertz (MC) (Table2). The radio waves produced by an alternating current vary in intensity with the frequency. While velocity and wavelength change as electromagnetic energy is propagated through different media, the frequency of a wave remains constant. It is therefore, a fundamental property.

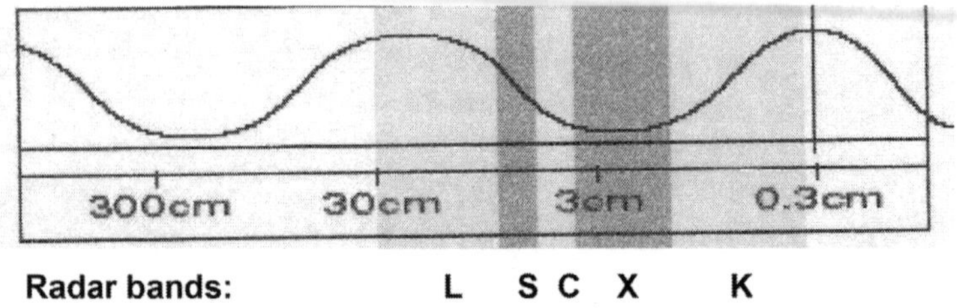

Fig.1.1 Microwave bands (After NASA)

Table 1: Electromagnetic spectral regions

Region	Wavelength	Usage
Radio	≥ 30 cm	Longest wavelength, Radio (sky and ground wave) communication.
Microwaves	0.3 to 30 cm	Longer wavelength passes through clouds, fog and rain, images in active or passive mode, radar communication.
Thermal IR band	3 to 5μm 8 to 14μm	Main atmospheric windows in the thermal regions. Images taken by optical mechanical scanner.
Reflected IR band	0.7 to 3.0μm	Photographic IR band using films.
Infra red	0.7 to 100 μm	Atmospheric transmission windows separated by absorption bands. Wavelength depending on interaction with matter.
Visible	0.4 to 0.7 μm	Images with films and photo detectors, clouds, earth and plants photography.
Photographic UV band	0.3 to 0.4 μm	Images with films and photo detector. Transmitted through atmosphere.
Ultraviolet	0.03 to 0.4 μm	Incoming rays less than 0.3 μm completely absorbed by ozone in the upper atmosphere.

Table 1: Continued

Region	Wavelength	Usage
X-ray	0.03 to 3.0 µm	Completely absorbed in the atmosphere, widely used in medical diagnosis.
Gamma rays	≤0.03 µm	Completely absorbed by the upper atmosphere, used in medical treatments.

Table 2: Terms used to designate frequencies

Units	Symbols	Frequency cycles per second
Hertz	Hz	1
Kilohertz	KHz	10^3
Megahertz	MHz	10^6
Gigahertz	GHz	10^9

If sun rays are passed through a glass prism, the light coming out from the prism will break into seven colours, each of a definite wavelength. This phenomena is known as spectrum. The term electromagnetic spectrum refers to a very wide range of electromagnetic waves arranged according to frequency or wavelength in continuous sequence. Microwaves are a band or a part of the frequency of the radio spectrum. They differ from radio waves due to difference in the technologies employed to access or record them. Still, the term radio frequency is sometimes used for microwaves, which is a misnomer. The microwave spectrum is usually assigned frequencies from 1 GHz to 100 GHz, although in the earliest days lower frequencies were included. The most common applications of microwaves are within 1 to 40 GHz range. Microwave frequency bands are generally abbreviated by letters such as P, L, S, C, X, K (Table 3).

Table 3: Microwaves Band, wavelength and frequencies

Band	Wave length (cm)	Frequency GHz (10^9 cycles per second)
Ka	0.8-1.1	40.0-20.5
K	1.1-1.6	20.5-18.0
Ku	1.7-2.4	18.0-12.5
X	2.4-3.8	12.5-8.0
C	3.8-7.5	8.0-4.0
S	7.5-15.0	4.0-2.0
L	15.0-30.0	2.0-1.0
P	30.0-100.0	1.0-0.3

Interaction of microwaves with matter

All forms of electromagnetic energy including microwaves that encounters or strike different substance of media (solid, liquid, gas) is called incident radiation. Its interaction with different media can change the following properties of the incident radiation.

- Intensity.
- Direction of travel.
- Wavelength.
- Polarization(plane of vibration).
- Phase (state of vibration).

Accordingly, incident radiation may be

i. Transmitted through the substance of different density. But it changes the velocity of electromagnetic radiation. The ratio of the velocities in to the velocity in the substance is called index of refraction or refractive index.

ii. Absorbed, giving up its energy which heats the matter.

iii. Emitted by the substance, usually at longer wavelength than the incident light. It depends upon the temperature and structure of the substance.
iv. Scattered i.e., deflected in all directions. The surface can scatter if the dimensions of particles or roughness of the surfaces have sizes similar to incident wavelength.
v. Reflected i.e., returned from the surface of a material with the angle of reflection equal and opposite to the angle of incidence. Smooth surface reflects most of the incident wave.
vi. Polarized i.e., the direction of vibration of reflected waves may differ from that of incident wave. This may occur in horizontal or vertical planes.

The basic physical principles of conservation of mass and energy hold good whenever electromagnetic waves interact with matter or a substance. Also, all matters radiate a range of electromagnetic energy with the peak of intensity shifting progressively toward shorter wavelength with its increasing temperature.

In addition, microwave have the following properties:

i. Microwave can detect objects in day as well as in night.
ii. Microwaves have the capability to penetrate haze, clouds and rain. Therefore, they can operate in all weather conditions. This, however, depends upon the frequency. A thumb rule is that S band, X band and K band microwave frequencies are best suited to detect rain, clouds and water drops (in clouds) respectively. Frequencies above 10GHz are attenuated by heavy rain.
iii. Microwaves penetrate more deeply into vegetation as compared to optical waves. The depth of penetration is more in dry soil or dry vegetation as compared to wet conditions. The penetration is less for higher frequencies.

iv. Microwaves are capable of penetrating into the earth depending upon the nature of surface and the moisture content. The extent of penetration depends upon the frequency of microwaves. Higher is the frequency, less is the penetration. However, sometimes higher frequency is needed to distinguish objects close by i.e., for better resolution.

v. Microwaves travel in straight line unless obstructed by hill, forests and other objects. They have therefore been used for land and satellite communications.

vi. Microwaves are non-ionising. They are not harmful like X-rays. Microwaves are therefore, preferred for diagnosis of tumors and other ailments, screening of public at the airports.

vii. The intensity (signal strength) of microwaves decreases withthe square of increase of distance (d) from the source. This property is similar to light and sound waves.

viii. The amount of heating of materials depends upon the power of the source which produces microwaves, their frequencies and duration

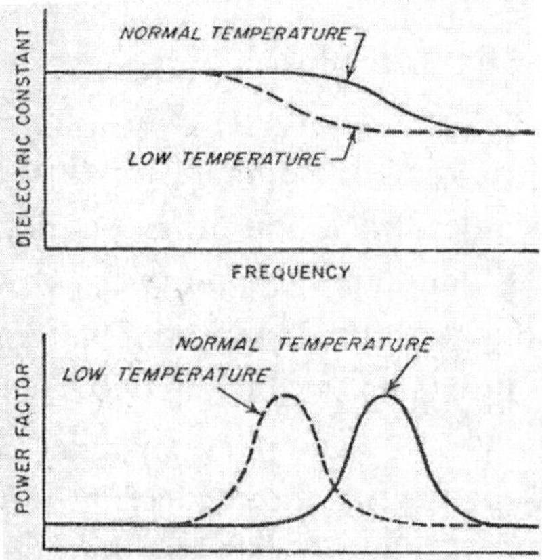

Fig.1.2 Variation of dielectric constant and power factor with frequency and temperature

Microwaves have some properties different from electricity (alternating current of frequency 50 Hz) used in our houses. Electricity flows through conductors or copper wires but it does not pass through insulators like wood. Some materials like resistances have low electric conductivity and are used to control the current in an electrical circuit. On the other hand, microwaves cannot penetrate conductors like metals; they are reflected from their surface. That is why if a metallic utensil is placed in a microwave oven, it stops functioning. Microwaves penetrate cement walls and roofs of our houses, thus enabling us to talk on mobile phones anywhere. In this connection, a term called dielectric material is used which is slightly different from insulators. Microwave propagation through a dielectric material is influenced by its electrical property called dielectric constant. When dielectrics are kept in an electric field, practically no current flows in them.This is because, unlike metals they have no loosely bound free electrons that may flow through the material. It is easy to remember that the capacity of a condenser is increased by inserting a dielectric material between its two plates.In practice, some loss of energy takes place when the capacitor discharges electrical energy. This is called power factor of the capacitor (Fig 1.2) and is attributed to dielectric losses. The dielectric material is made up of atoms and molecules. In these atoms, the negative charges called electrons surround positive charges in the nucleus called protons which are balanced to make them electrically neutral. When the material is placed in an electric field, its molecule is distorted. This is due to their shifting from their equilibrium position under the influence of electric field. The phenomenon is called dielectric polarization. The positive charges are therefore, displaced toward the field and negative charges shift in the opposite direction. An internal field is therefore, created which reduces the overall field within the dielectric (Fig.1.3).

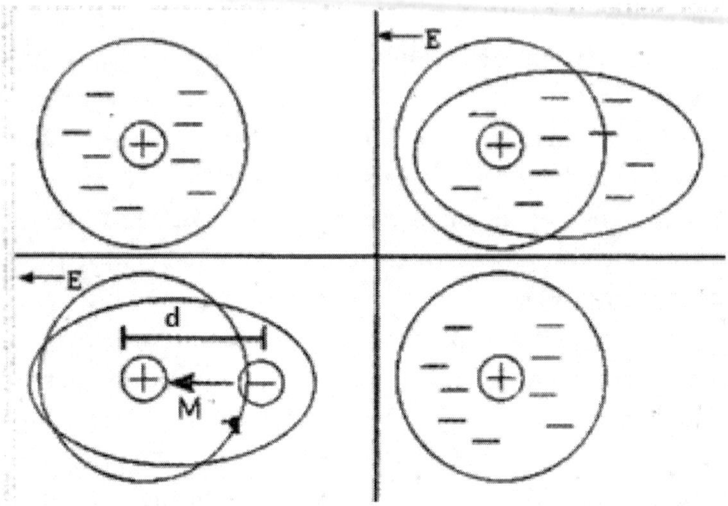

Fig. 1.3. Electric field interaction with atom of a dielectric (M is the dipole moment and E is the electric field)

Let us now take the example of a molecule like water. It consists of two atoms of oxygen and one atom of hydrogen. When microwaves are passed through water, its molecules will start rotating which collide with other molecules due the phenomenon of polarization as explained above. It is measured by a quantity called dipole moment (electric charge multiplied by the distance) which orients generally in the same direction as electric field. After the removal of electric field, the molecules of the material return to its original state. The time lag when they return to original state is called relaxation time. Alternately, the average time taken by the molecules to become random after switching off the electric field is called relaxation time. It differs for different substances depending upon

i. The size of the molecule in the substance or its shape.
ii. Viscosity (fluids – gases or liquids).

These properties are inherent in the electrical quantity called dielectric constant of the material. The dielectric property is different for various substances like air, water, earth, soil and rocks and changes with the temperature. Their dielectric constant also depends upon the frequency of electromagnetic waves. This phenomenon is called dispersion. When the frequency is very high as in the case of microwaves, some part of its energy is converted into heat and the rest passes through the medium. The energy loss or dissipation factor is called loss tangent. It is therefore, a measure of the ability to absorb microwave energy and conversion into thermal energy or heat. Therefore, at high frequencies, the definition of dielectric constant is modified to include both these properties. Mathematically, it is defined by a complex quantity with real part and imaginary part (dielectric loss). In short, the dielectric properties are important to understand the storage and dissipation of electric and magnetic energy in materials. The real part in complex dielectric constant causes bending or refraction as it passes through the medium. The complex part gives the attenuation of radiation as it passes through an absorptive medium. At microwave frequencies, the dielectric constant of dry rocks and soil ranges from 3 to 8. But the propagation of microwaves which are most effected in atmosphere, ocean and earth are attributed to the presence of water whose dielectric constant is 80. Thus, if the moisture content of a material increases, the dielectric constant of the substances increases. In the atmosphere, the moisture occurs as a gas in the form of water vapour. The dielectric constant of moist air is slightly larger than dry air whose dielectric constant is 1.

All the dielectric substances can be

i. Polar which have dipole moment, dielectric loss and relaxation time like water.
ii. Non polar without the above properties like benzene.

To summarise, the polarization phenomenon enables a molecule to rotate very fast in the microwave region. The time lag in response to change of electric field causes heat and is used in microwave ovens and medical therapy.

CHAPTER 2

Atmospheric effects on microwaves

Most of the applications of microwaves used in satellite communication, radar, GPS or missile control depend upon their passage in the atmosphere whose composition affects their propagation.Microwaves travel in straight line, but bend in the atmosphere depending upon the weather conditions. This phenomenon is described by a term called refraction which means a change in the original direction of the path of wave. This can happen if the medium, say atmosphere is not homogenous or if the wave meets the boundary of two different media. When a ray travels from denser to rarer medium, it is bent away from the normal (Fig 2.1).We know that the atmosphere becomes less dense as we go higher. In other words, the density of the atmosphere decreases with height. This also results in decrease of refractive index with height as it is dependent upon the atmospheric pressure, temperature and humidity or water vapour. The effect is that if a radio ray is sent upwards, it will start bending towards the earth due to lower atmospheric pressure (density) or refractive index in higher layers. If the atmospheric bending becomes so large, that the radio ray strikes back the surface of the earth, it will be reflected back into the atmosphere. Under certain weather conditions, it may again be bent so much that it is reflected repeatedly between earth's surface and top of such layer as in a wave guide. This phenomenon is called super refraction or ducting(Fig.2.2). The layer of the atmosphere in which repeated reflections between its top and the ground surface occurs, is called duct. The distance travelled by microwave signal in a duct will, therefore, be increased far beyond that in normal weather conditions.

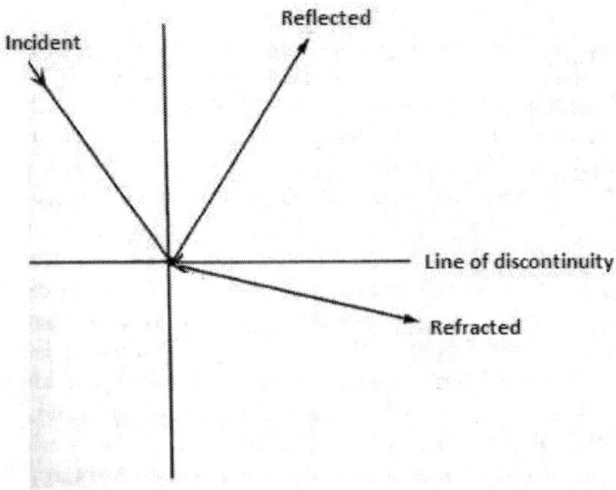

Denser medium

Fig. 2.1 Reflection and refraction phenomenon

Structure of the atmosphere and microwave time delay

The structure of the atmosphere plays an important role in microwave propagation.It is divided into different layers (Fig. 2.2).

(i) Troposphere is the lowest layer in the atmosphere closest to earth. It has average height of 10 to 12 Km varying from about 8 Km near the polar region of the earth to 16 Km over the equator. The weather mainly occurs in this layer. The atmospheric pressure, temperature and humidity (water vapour in gas form) generally decrease with height. The lower part of the atmosphere has relatively higher amount of water vapour or humidity and causes delay in travel time of microwaves. This is attributed to lower velocity of electromagnetic waves in moist air as compared to dry air.

(ii) Stratosphere:
In this layer, the temperature remains constant or slightly increases. It extends upto a height of 50 Km in the atmosphere.

(iii) Mesosphere:
The temperature again decreases slowly in this layer extending upto height of 85 Km or so; thereafter temperature increases continuously in the thermosphere.

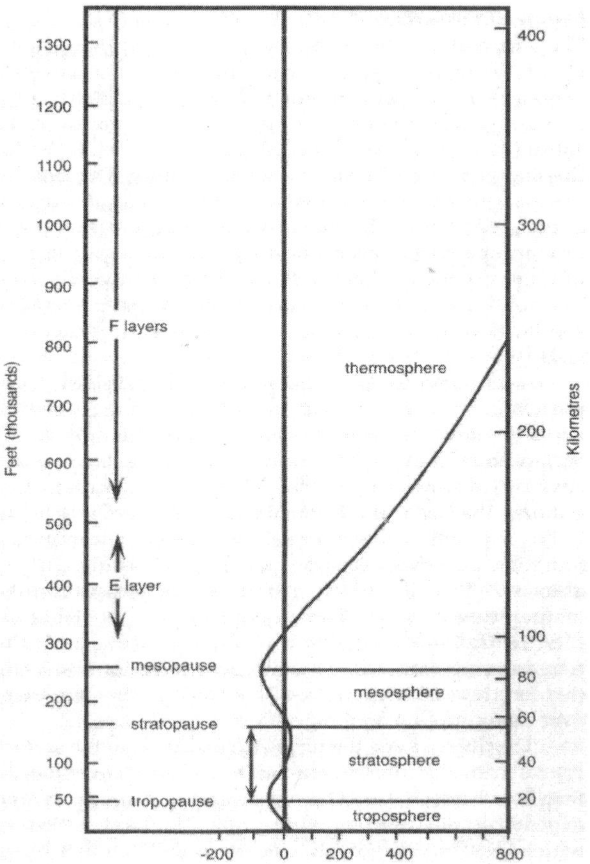

Fig. 2.2 The atmosphere

(iv) Ionosphere:

Energy radiated in direction above the horizon will travel through space until it reaches the ionized region in the upper atmosphere. This part of the atmosphere where ionization is appreciable is called ionosphere. This region absorbs large quantities of radiant energy from the sun. It not only gets heated but also produces some ionization in the form of free electrons, positive and negative ions. The ionosphere is divided into three layers called D, E and F layers. The layer D extends between 50 and 90 Km. The height of E and F layers are generally 110 and 220 Km respectively. During day time, the layer F breaks into two layers called $F_1$and F_2 which merge into a single layer F_2 during the night time. The E layer ionization is attributed to ultraviolet light from the sun. It decreases uniformly with time during the night. The D region ionization is largely absent at night.

The effect of high density of electrons reduces the dielectric constant or refractive index of the layer. This helps in the bending of radio waves towards the earth. The long distance radio communication is due to the sky wave being reflected from the E and F layers of atmosphere. However, if the frequency of the wave is higher, say in the microwave range, its velocity will be reduced in the region of higher electron density due to decrease of refractive index. Thus it will take a longer time to travel same distance in the atmosphere. This is called delay time. Corrections have to be applied to microwaves for troposphere and ionosphere delay times as required in Global Positioning Systems (GPS) to obtain precise location of objects through satellites.

As mentioned earlier, the extent of bending (refraction) is given by the refractive index of a medium. It is practically the same for radio and light air. The refractive index of a gas is, therefore, effectively equal to the square root of the dielectric constant similar to the case of light waves. But its value changes in atmosphere because in addition to oxygen, carbon dioxide, trace gases, it contains water vapour (in gaseous form) called humidity. The water molecule responds differently to radio waves of different frequencies. The dielectric constant of water is 80 for radio and light waves but it decreases for microwaves. Therefore, for propagation of microwaves in the atmosphere, its refractive index commonly called radio refractive index at different heights is required. The radio refractive index in atmosphere consists of a dry part and a wet part. The radio refractive index being quite small is replaced by a large number by multiplying it by 10^6 and including a term for the effects of curvature of the earth. This is called modified refractive index. The need for this arises due to the fact that microwave propagation involves very long distances and therefore curvature of the earth surface needs to be taken into consideration.

Microwave propagation problems can be simplified by considering ray paths as straight lines instead of being curved. The wave propagation can be represented geometrically by straight lines showing its direction of travel. It is more common to describe the waves as rays just as in the case of light (Fig.2.3).

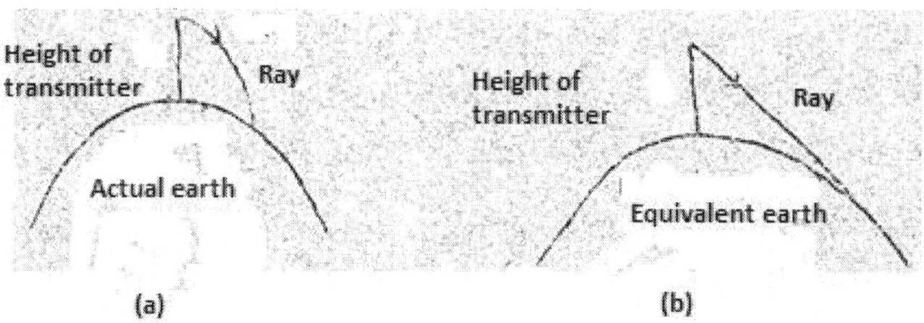

Fig. 2.3 Rays transmitted in atmosphere over a) actual and b) equivalent earth

In the atmosphere, the curvature of the ray is towards more dense or its lower region. The curvature of the earth is expressed as the reciprocal of its radius. The total curvature of the ray downwards towards the earth can be calculated by adding the curvature of the earth. For tracing of rays in an atmosphere so that they can be represented as straight lines, the modified radius of earth is taken. The curvature of the rays is compensated by increasing the radius of the earth. Where the curvature of effective earth radius becomes zero, the rays are straight lines. The distance of the line of sight or optical horizon is equal to the square root of 2 times the height of the radio transmitting antenna above the earth's surface and its radius. The distance of the radio horizon is increased by about twenty percent due to the effective earth radius in the atmosphere. To get an idea of ducting conditions, generally the decrease of humidity in excess of ½ g/ Kg per 30 meters and temperature inversion in excess of about 6 ^{0}c per 30 meters cause downward bending of radio rays.

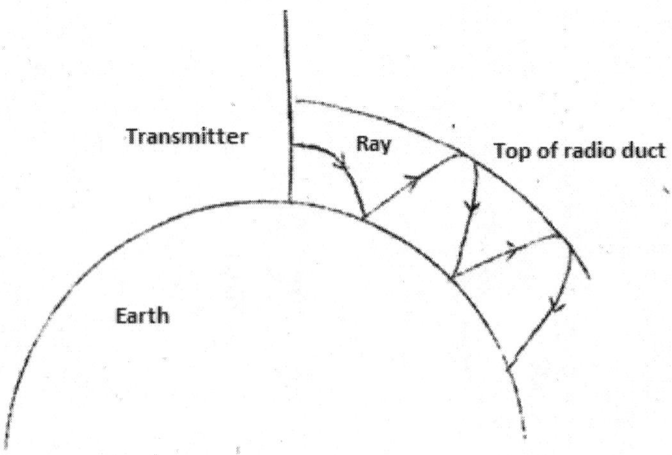

Fig. 2.4 Ducting phenomenon in atmosphere

It is now easy to understand that when duct propagation exists, the energy or microwave signals will travel to great distance around the curved surface of the earth. The attenuation or loss of these signals will be low as the energy is trapped in the duct as it travels in a series of hops due to successive earth reflection(Fig 2.4). This condition occurs when the humidity at the surface of the ground is very high and decreases rapidly with increasing height. Also, the atmospheric temperature increases rapidly with height as in a temperature inversion during winter. Thus in terms of modified refractive index, bending of microwaves is governed as follows:

(i) If the modified refractive index does not vary with height, the bending of waves will be same as that of the earth. This happens at the top of the duct.

(ii) If the modified refractive index increases with height, radio waves will be bent away from the earth.

(iii) If the modified refractive index decreases with height, radio waves will be bent towards the earth. This situation facilitates ducting conditions.

Due to large number of weather recording stations (weather observatories), the observations of atmospheric pressure, temperature and humidity are readily available. Hence it is easier to compute microwave refractive index from weather observations.

Influence of weather on microwave propagation therefore, gives rise to super refraction, subrefraction phenomena or fading of microwave signals. During WorldWar II, sometimes, the radar at Mumbai coast could locate ships as far as Aden coast in Arabia. This means that the range of a radar which is usually 200 Km may be increased to thousands of km or more due to super refraction phenomenon. During winter, the super refraction phenomenon occurs commonly in the morning due to increase of temperature with height called temperature inversion. On the in land stations the meteorological conditions cause variations in the received signal. This is due to the changes in radio refractive index which depend upon three meteorological factors namely atmospheric temperature, atmospheric pressure and humidity.Vertical gradients of temperature and humidity in the atmosphere may be modified in many ways to cause favourable conditions for super refraction phenomenon or duct formation. Heat or moisture may be increased or decreased from the air in contact with the earth's surface in several ways. There is generally heating during day time and cooling during night. Evaporation from the surface or water bodies like ponds, rivers and oceans may increase water vapour or decrease it by the formation of dew. The microwave refractive index variations in the atmosphere depend upon:

i. Subsidence or sinking of air such as in areas of high atmospheric pressure called anticyclones which can modify the vertical temperature and humidity structure of the air. It frequently happens in the centre of stagnant anticyclones (high pressure areas) that large layers of air sink, or subside, towards the ground. This sinking is accompanied by an outflow of air from the area, horizontally near the surface since it is subjected to larger pressure in the lower levels close to earth's surface. It is compressed consequently, being confined in a small pocket where the air temperature rises. Under certain conditions, the decrease of temperature with height called the lapse rate decreases and may change to an inversion causing increase in temperature with height. It is also seen that subsidence or sinking of air will generally lead to a more rapid decrease of humidity with height. If air sinks in an anticyclone, the low value of the humidity corresponding to higher levels will be brought nearer to ground. This will result in increasing the vertical gradient of humidity.

ii. Advection: On horizontal surface of the earth, the movement of air is caused by the wind from one region to another region where the surface temperature or the vapour pressure is markedly different. For example, the air may be originally over a land surface, where it has been warmed and may then be carried across the coast and pass over a comparatively cold sea. The surface layers of the air are thus cooled and an inversion is formed.

iii. Eddy diffusion: Small fluctuations in the air gives rise to turbulence described by eddies. High eddy diffusion means almost uniform temperature and humidity and consequently normal propagation. Low eddy diffusion permits the existence of substantial atmospheric gradients and so makes super refraction possible. It may be mentioned that eddy diffusion is low in temperature inversion.

iv. Sea breeze: Sea breeze occurs in the afternoon due to difference in the amount of sun's heat received over coastal regions and land. An effect of the sea breeze is to bring moisture inland in the lower layers for a short distance, creating conditions favourable for the formation of a duct. Although it is limited to an area of a few kilometers of coast line, it may be useful for coast line microwave transmission.

Nature of the Surface.

Since microwaves travel in straight line, any obstruction in their path reduces signal. Therefore, nature of terrain is important. The sea presents an essentially flat surface, as the waves are too small to interfere with the radio propagation. Consequently main factors for the formation of a duct over the sea are attributed to meteorological factors. Similarly, over desert areas, the conditions may be almost uniform but over the land the undulations of the surface and the variations in type of soil, vegetation, forests make it difficult to form a duct over an extensive area. In forests and hilly terrain, duct formation becomes more complex.

Thunderstorm super refraction

An important source for ground ducts is the diverging down draft under a thunderstorm which causes super refraction phenomenon. When the relatively cool air spreads out from these thunderstorms, it results in a temperature inversion in the lowest layers close to earth. The air, of course, is humid which favours a strong duct over relatively small area in the vicinity of the thunderstorm. From careful radar observations, one can detect the establishment of anomalous propagation conditions by the sudden increase in the number and range of ground structures on the radar screen.

Effect of fog

Fog may be associated with either super refraction, sub-refraction or with a fluctuating signal strength. Within the fog layer, the vertical gradient of neither temperature nor humidity is likely to be sufficiently large to cause super refraction. Within this layer, therefore, propagation will be normal or even sub normal. Above the fog layer, however, there may be a noticeable decrease of humidity with height leading to the formation of an elevated duct. Fogs which occur on clear nights are frequently shallow, in the early stages, and either the transmitter or receiver or both may be in the fog layer. Any duct which exists above the fog may or may not be sufficiently long to cause anomalous propagation.

Diurnal and seasonal variations of surface refractive index

The downward bending effects of microwaves are caused by the gradient of radio refractive index in the atmosphere. Largest bending of rays upto 70% are found to be largely effected by surface values of radio refractive index or its gradient in the lowest one kilometer.

For the Indian region, the decrease of refractive index with height has been expressed by models called exponential model (in which surface refractivity decreases exponentially with height), biexponential models (in which both the dry and wet terms of surface refractive index vary exponentially with height), or a few other models. Use of these models can improve the accuracy of Global Positioning System (GPS), particularly in the vertical direction (discussed in Chapter8).

The diurnal variations of microwave refractive index are utilized to estimate the transmission loss or decrease in the signal strengths. In general, this index is maximum in the morning during 00 to 03 hours and minimum in the afternoon during 09 to 12 hours. The signal strengths of a receiver may therefore, be expected to show similar diurnal trend. The diurnal changes are largest during pre monsoon season

Weather systems in the tropical regions

The weather conditions in India rapidly change with the low pressure weather systems which move over northwest India or develop over oceanic areas in Bay of Bengal and Arabian sea. Due to marked change in temperature, humidity and rainfall, they affect microwave propagation.

i. **Western Disturbances:**
ii. The low pressure systems which generally originate near the Mediterranean sea, Caspian sea and Black sea move westward. These low pressure areas are called western disturbances which give rain or thunder showers over northern India mainly during winter season. The change of the value of the microwave refractive index during a

western disturbance is of the order of +45 N_O units with the approach of the system which reduces to about 25 N_O units in the rear i.e., after it passes away. These values, however change depending upon the intensity of the system. After the passage of weather disturbances, cold air flows over northern India. This favours the formation of temperature inversion and super refraction phenomenon in winter.

iii. **Monsoon depressions and cyclonic storms**

Monsoon depressions are also low pressure systems which form during monsoon period generally over Bay of Bengal and sometimes over Arabian sea when the fall of .pressure is at least 4 hectopascals (mb) or larger. They move over Gangetic plains causing heavy rains. A few depressions may also form in Arabian sea which may strike Maharashtra and Gujarat coasts. The depressions cause almost continuous rainfall for two or three days. Highest rainfall occurs in the southwest sector of the monsoon depressions. The pattern of rainfall has significant effect on microwave propagation. The average value of the 24-hourly rise in surface refractivity based on several monsoon depressions at surface, 1.5 Km, 3Km, and 6Km, mb levels may be of the order of 08, 16, 08 and 05N units respectively.

The more destructive weather system called cyclonic storms occur during premonsoon (March to May) or post monsoon (October, November). Larger number of these storms originate in Bay of Bengal but a few also occur in Arabian Sea, and cause destruction in the coastal regions due to strong winds, storm surges, heavy rain and flooding. Navigation system fails if a ship is caught in the cyclonic storm.These cyclones over the Bay of Bengal and Arabian Sea originate as low pressure areas

(LPA) over the south Andaman Sea, south Bay of Bengal and southeast Arabian Sea during the cyclone seasons (namely April, May or September to November). Sometimes they move from the east as the remnants of typhoons in the Pacific ocean and intensify when favourable temperature of about 28^{0}C occurs over oceans.

Under certain meteorological conditions over ocean, super refraction phenomenon may occur. There are two different air masses present in the summer monsoon circulation over the Arabian Sea. The lower air mass, (cool and moist of maritime origin) after crossing the equator in the extreme western Indian Ocean enters the Arabian Sea from south and the overlying air mass which is hot and dry (of continental origin) enters in the region through the western boundary. These air masses get transformed towards the west coast of India after moving across the Arabian Sea. The observed refractive index variations over the region can thus be explained in the light of these meteorological conditions which favour the formation of temperature inversions and the humidity gradients over the region near west coast of peninsular India.The gradient of radio refractive index in the lowest one kilometer during June and July is of the order of 55 to 60 units at Bombay. This value decreases near the latitude of 11^{0}N. It is therefore, inferred that relatively better super refracting conditions occur over north-west Arabian Sea as compared to the southern region of west coast of India as found from Indo Soviet Monsoon experiment.

Duct heights are generally a few tens or hundreds of meters. Accordingly, duct propagation is limited to ultra high frequencies or microwaves. It may be noted that if transmitter height is greater than the duct height, its effect is negligible on the signal strength of microwaves. On the other hand, the transmitter height within the duct will cause large signal strength over long distance in favourable conditions.

Troposphere scattering

Tropospheric scattering or topo scatter is the name given to a different kind of extended range propagation other than super refraction phenomenon. This is attributed to scattering and reflection of microwave caused by

i. Bending or diffraction around the curved surface of the earth in the same way as sound waves which bend around a corner.
ii. Turbulence or blobs in the troposphere which give rise to small irregularities in the refractive index.

This extended range propagation is largely independent of frequency of microwaves. The use of toposcatter method is advantageous for microwave communication link over islands(Fig2.5).

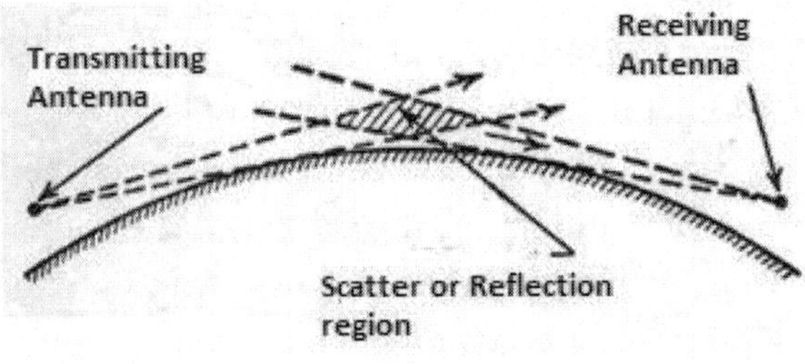

Fig. 2.5

Fading

As discussed earlier, the influence of the changes in meteorological factors and consequently radio refractive index may result in loss of signal strength from the transmitter to the receiver. Efforts have been made to take some remedial measures as follows:

Under, sub-standard atmospheric refraction such as movement of vapour laden fog over cold sea water, there is a sub-standard increase of path loss over a wide frequency band. Such fading can be reduced by increasing the height of the antenna.

i. Interference between two or more rays in the atmosphere gives rise to microwave fading on path with adequate clearance. The separate paths between transmitter and receiver are caused by irregularities in the variations of the refractive index with height. The overall system performance is determined by providing transmission margins for the loss in signals.
ii. Super refractive atmosphere, particularly at night increases the range of the transmitter but causes fading due to exposure to ground reflections. Misaligned antenna cause night time fadings. It is, therefore, necessary that larger, narrow beam antenna should be precisely aligned to track the shift of the microwave ray during these periods.
iii. Multipath fading occurs during night time particularly when the microwave paths are quite long. During the day, the lower atmosphere is thoroughly mixed by convection caused by sun's heating and wind. This results in steady signals on the line of sight paths. But on cloudless nights, with little or calm wind, sizable irregularities or layers may form. These irregularities result in localized changes in the refractive index and result in bending of microwaves and multipath transmission. Multipath fading therefore may build up during night and reaches peak value in the morning. These super refractive layers are broken in the day by convection caused by heating of earth due to sun. Multipath fading may also result from the elevated atmospheric layer in some cases. The effects of multipath fading can be minimized by the proper use of either frequency or space diversity.

CHAPTER 3

Microwaves generation, transmission and reception

Microwaves used in different appliances like oven, radar and others are generated just like electricity. Different amount of power is required for specific purpose and accordingly different type of microwave generators are used. For sending them over long distances, special types of antenna systems are made. After their reception by microwave receivers they can be recorded and used in desired application. However, in applications like oven or medical therapy and diagnosis or laboratory instrumentation, antenna is not required due to short distance. In these applications guided wave propagation mechanism is used.

Microwave generators

The limitations of vacuum tubes like triode and pentode arise at very high frequencies due to time taken by electrons to travel from cathode to anode. But this transit time of electrons is used to design klystrons and magnetrons for generation of microwaves.

(i) **Klystrons**

(ii) Just as electron tubes have been used for amplifying or generating radio frequencies, klystrons are special type of vacuum tubes specifically developed for microwave frequencies. They are used as amplifiers to produce low power reference signals for super heterodyne radar receivers (described later). Also, they produce high power carrier waves to carry sound, video and data as required for microwave communications. Other applications include linear accelerators and industrial heating. Their frequencies range from 300 MHz to 40 GHz with peak power output of 100W to 150 MW. The functioning of klystrons, is easier to understand if we remember the following :

Producing a beam of electrons by a cathode of special material which emits electrons when heated.

i. Accelerating the electrons beam produces a radio frequency wave.

ii. Applying an alternating electric field which accelerates or slows down the motion of electrons in such way that they form bunches. Bunches can be understood by throwing a ball upwards. It will reach ground after some time. A ball thrown upwards with greater velocity will take more time to return to the same distance on the earth as the first. Thus by velocity modulation, the balls can be made to strike at the same point. Likewise electrons also cause bunching.

iii. **Power grid tubes**

Like a triode, it consists of a cathode, grid and anode. Between the gird and the cathode, the weak microwave signal is fed. This signal to the grid controls the electrons flow from the cathode and amplifies the signal at the anode. These are used for producing frequencies of 350 MHz upto 1KW with the maximum power of 2MW at 30MHz.

Traveling wave tube

The travelling wave tube is an amplifier that is particularly suitable for amplification of very high frequencies such as 3000MHz and higher. It consists of Electron gun which produces pencil like beam of electrons. High voltage of about 1500 volts accelerates the electrons to produce pencil like beam.

i. Loosely wound helix through which the beam is sent
ii. Anode collects the electron beam.
iii. An axial magnetic focusing field to guide it through the centre of the helix.
iv. The weak signal to be amplified is applied to the end of helix adjacent to the electron gun. The amplified signal appears at the other end of the helix. The input and output coupling may be co-axial type or wave guide. Similar to a klystron, the electrons bunch in a travelling wave tube which enables them to deliver energy to the wave. The backward wave oscillations are prevented by introducing an attenuator near the input end of the tube. It absorbs any wave propagated along the helix. The amplification of the tube is constant to within ±3 db from 2000 to 4000 MHZ.

Magnetron

This is most commonly used microwave generator. It was the basis of radar transmitters in World War-II and is still being used widely. A magnetron converts electrical energy into microwave radiation. These are of two types namely,

i. Pulsed magnetron which produces output power from kilowatts to several megawatts for very short spells (duration)
ii. Continuous wave magnetron produces continuous output power in the range of a few watts to kilowatts.

Design and operation

The essential parts of magnetron oscillator are :-

i. Cylindrical Cathode
ii. Anode structure having cavities opening into the cathode – anode or interaction space by means of slots. (Fig 3.1)
iii. Tapered wave guide or a coupling loop to withdraw power.
iv. Permanent magnet to provide external magnetic field with flux lines parallel to the axis of cathode.

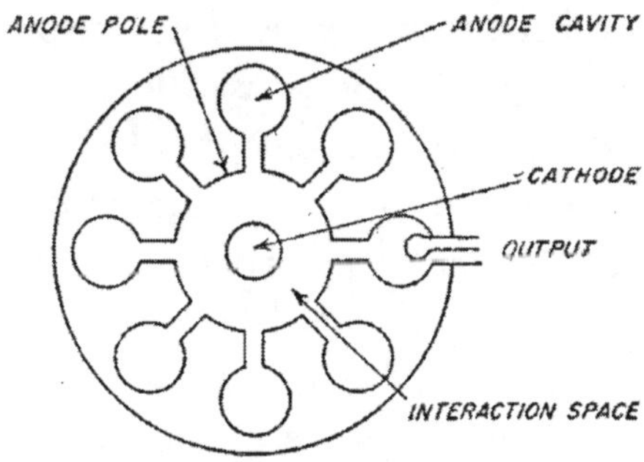

Fig. 3.1 Cross section of cavities of magnetron

In a magnetron, electrons are emitted from the hot cathode. Both the magnetic and electric fields are applied which are perpendicular to each other. They make electrons to rotate in circular direction in the coaxial space between the cathode and the anode under the influence of an axial magnetic field. The electromagnetic field causes alternating positive and negative voltages on the anode. The acceleration or deceleration fields produce bunching effect on electrons forming an electron cloud. The focusing action is equivalent to a velocity modulation, which causes bunching of electrons. It causes the orbit of electrons to be confined to spokes, one for each two anodes. In the case of p mode, these spokes rotate at an angular velocity corresponding to two poles per cycle. A certain fraction of electrons emitted from the cathode travel out through the spokes. They deliver energy continuously to the oscillations until the electrons reach anode and disappear. Electrons emitted in the portion of the cathode between spokes are returned very quickly to the cathode and in this process absorb only a small fraction of the energy. The net result is that the oscillations receive a substantial net energy from the electrons.The resonant system in a magnetron has several resonant frequencies which are more commonly called modes which are equal in number to the number of cavities. These cavities in the resonant system give rise to resonant frequencies as a result of coupling. The mode used in a magnetron is such that the phase difference between the adjacent anode poles is p radians. This is called p mode. Magnetrons are useful for generating high powers at frequencies in the range of 1000 kw to 25000 MHz. Peak power exceeding 1000 kw can be obtained in magnetrons designed for pulsed operation at 3000 MHz. It gives 1 kilowatt continuous wave power obtained at the same frequencies. These have efficiencies of 30 to 60%. High power magnetrons are cooled by water.If we strike a ball once, sound wave of single tone (frequency) is heard. If we strike it many times, same frequency will be heard louder. Similarly in magnetron, the power output depends upon pulse repetition frequency.

Gyrotrons

Gyrotrons are high powered electron tubes which emit a millimeter wave beam by bunching electrons with cyclotron motion in a strong magnetic field. They have been designed to produce higher powers (kilowatts to several megawatts) at millimeter wavelengths as compared to other tubes. The frequencies at which they operate are 8 to 800 GHz.Gyrotrons can be designed for pulse or continuous operation . They are mostly used in fusion reactors. Cyclotron resonance heating of fusion plasma in these reactors is used for sintering of ceramics and joining of metals. (The word plasma is used to denote assembly of charged particles. It is formed when the atoms in a gas are ionized to produce equal number of ions and electrons for example in a gas discharge tube).

Modern microwave oscillators utilize a solid state device like diode or transistor, a resonant circuit and a matching network to convert DC power to microwave power at the designed frequency. This method is used in miniature microwave gadgets like a cell phone.

Microwave antenna

Radio waves carrying voice, sound or data do not require wires or cables as used in telephones. The transmission of radio or microwave signal is done by antennas. They are received at different places by receiving antennas. The transmission and reception of signals can be done by the same antenna as in radar. The laws of electromagnetic waves are governed by the famous Maxwell's equations named after their discoverer. These waves show the properties of magnetic as well as electric field in electromagnetic wave. It is also found that the strength of the radiated field decreases inversely with distance.

It may be interesting to note that every electrical circuit which carries alternating current radiates a little electrical energy in the form of electromagnetic waves. However, if the dimensions of the circuit are nearly equal to the magnitude of a wave length, considerable amount of energy is radiated. The size of the radiator is inversely proportional to the frequency. High frequency waves can therefore be produced by a small radiator (dish antenna) in contrast to very large antenna system required by low frequency waves.

Every radiator sends out stronger waves in some directions as compared to other direction. This property is used to transmit or receive energy towards or from a particular direction.

The strength of a radio wave is measured in terms of voltage produced in space by the electric field of wave. It is expressed in microvolts per meter. Due to variation in intensity caused by alternating current (called sinusoidal or sine wave type variation), the average value of 0.707 times the maximum voltage is taken as the intensity.

Radio antennas have the following properties:

i. They radiate the radio frequency energy that is generated in the transmitter and guided to the antenna by the transmission line.
ii. The antenna directs energy into desired directions.
iii. The radiations in other directions are suppressed. The radiation is expressed as field strength E volt force per meter. It gives electric field. If the radiation is expressed in a given direction in terms of power solid angles, it defines power pattern.

Dish Antenna (portable antenna)

With the advent of mobile phones and television, dish or parabolic antennas are seen in most of the places. Its curved surface reflects the radio waves (microwaves) just like a search light reflector. It directs these waves in narrow beam. Since at microwave frequencies wave lengths are small, convenient size of dishes are used to transmit or receive them(Fig 3.2). Different methods to focus energy are shown in Fig 3.3 .

Fig. 3.2 Dish Antenna

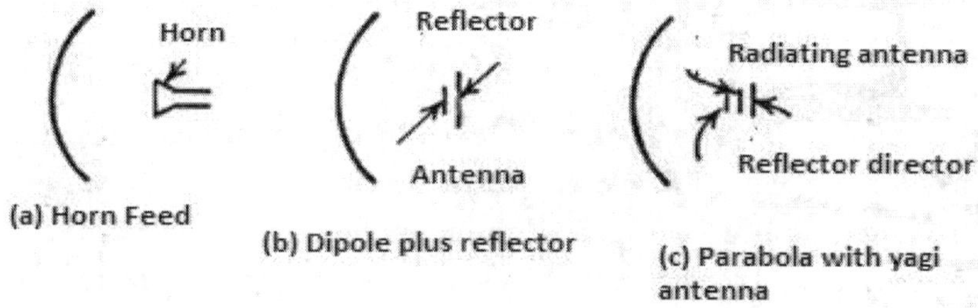

Fig. 3.3 Feed antenna to focus microwaves

The dish antennas are used as high gain antennas for point to point communications such as microwave relay links for telephone and television between nearby places. Wireless WAN/ LAN links are used for data communication, satellite and spacecraft antennas and radio telescopes. Their utility in radar antennas lies due to narrow beam of such waves which is useful for locating clouds, aircrafts and other objects. The concept of developing parabolic reflectors for radio antenna was taken from optics (light) where a parabolic mirror is used to focus light into a beam.

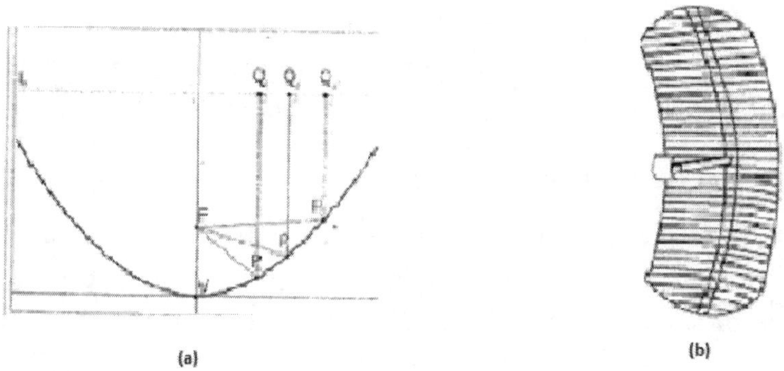

Fig. 3.4 (a) Parallel rays Q after striking parabolic surface at point P are focused at F (b) Wire mesh parabolic antenna

The principle of a parabolic antenna can be understood by comparison with a parabolic reflector placed in front of light waves. The source of micro waves is placed at the focal point (focus) in front of a parabolic reflector of conductive materials. It will be reflected from the concave surface into a plane wave in the form of a beam. Similarly, an incoming plane wave parallel to the axis after striking parabolic surface (antenna) will be focused to a point which lies at the focal point or focus (Fig 3.4a).

The design of a parabolic antenna includes another small antenna called feed antenna which is placed at the focus with its face towards the dish reflector(Fig 3.3). The parabolic reflector is made of metal and truncated in a circular rim that forms the diameter of the antenna. A transmitter which generates microwaves is connected by transmission cable to the feed antenna. These waves are emitted by feed antenna towards the dish antenna which reflects the waves into a parallel beam. It travels in straight line till it is intercepted by an object. These waves are then reflected or scattered by aircrafts, clouds, birds or any other object in the path of the rays. These are received back by the dish antenna. This antenna bounces the radio waves to focus to a point at the feed antenna. It is then converted into electric currents which travel through a transmission line to the receiver. After amplification, they are recorded by a digital recorder. Their visual display enables to distinguish objects in the sky.

Different designs of dish antennas

If we observe several cell towers, we will notice that a large number of dish antennas differ in design. These reflectors are made of sheets metal or wire grill construction(Fig 3.3b). They are either a circular dish or have other shape to create different beam shapes.

A metal screen works like a solid metal antenna. The holes in the mesh are smaller than 1/10 of a wavelength otherwise energy will escape causing loss of signal. These have an advantage of being lighter and have lesser effects of strong wind. The shape of the dish is designed to be accurate within a small fraction of a wave length. This ensures that the waves from different parts of the antenna reach at the focus in phase. An antenna made of grill of parallel wires or bars oriented in one direction acts as a reflector as well as a filter. It helps to filter out noise and improves quality of signal(Fig.3.3b).

Dish or parabolic antenna is more common and is shaped like a parabolic truncated shape in a circular rim. Along its axis, a narrow pencil shape beam is radiated.A distinct advantage of parabolic antennas is that they can function over a wide range of frequencies. They have wideband width and are non – resonant.

The other designs include:

i. **Shrouded dish:** A cylindrical metal shield is attached to the rim of the dish called shrouded dish. It shields the antenna from radiation coming from outside the main axis beam. Thus side lobes are reduced which give best signal to noise. It is more useful in terrestrial microwave links where interference is reduced from several antennas using the same frequency located together. The shroud is also coated inside with microwave absorbent material.
ii. **Shaped beam antenna:** Modern parabolic antennas are designed to produce a beam of specific shape instead of narrow pencil beam of the simple dish antenna. The shape of the beam is controlled by different curvatures in the horizontal and vertical direction. The advantage is that wider antenna in a transverse direction gives narrow radiation pattern in that direction. Such antennas are generally used in radars.
iii. **Orange peel antenna:** It is a long narrow antenna shaped like the letter, 'C'. It radiates a narrow vertical fan shaped beam.

It is often used in air surveillance by military to measure altitudes of objects by radar.

iv. **Horn or feed antenna:** The feed antenna is kept at the focus of parabolic antenna. In some cases, a secondary reflector is needed to divert microwaves into the parabolic reflector and is placed away from the feed antenna. Some parabolic antennas transmit or receive at multiple frequencies by having several feed antennas mounted at the focal point and are placed close together. The feed antenna is connected to the radio frequency equipment for reception or transmission by coaxial transmission line or wave guide. The frequency of operation of parabolic antenna is changed by the feed antenna which works at that frequency.

The classification of parabolic antenna is sometimes based on the method of supplying microwaves to the antenna through horn or a feed antenna:-

i. **Axial or front feed:** The feed antenna is placed in front of the parabolic dish at its focus on the beam axis. This is most commonly used. The only disadvantage is that the feed and its support block some portion of the beam. This reduces the aperture frequency to only 55-60%.

ii. **Offset feed:** This design enables to shift the feed structure out of the beam path to avoid blocking the beam. The focus and the feed antenna are kept out at one side of the dish. This method is widely used in satellite television dishes.

iii. **Cassgrain feed:** In this, the feed antenna is located on or behind the dish. It radiates forward and illuminates a convex hyperboloid shaped secondary reflector at the focus of the dish and is then reflected from the dish as outgoing beam. This method is used in large satellite communication antennas and radio telescopes. Its aperture efficiency is higher of the order of 65 to 70%.

iv. **Gregorian feed:** This is similar to Cassegrain design with the difference that the secondary reflector is concave in shape. Its aperture efficiency is more than 70%.

Gain of antennas

An antenna concentrates its radiated energy which in some cases to certain preferred directions. These directive qualities are measured by its gain. The extent of such concentration relative to that of some standard antenna defines gain. The standard isotropic antenna produces waves that are of equal strength in all directions. Although an isotropic antenna does not exist, the properties of such an imaginary antenna are easily visualized. The gain of an antenna is defined as the ratio of the power received by the antenna from a source along its beam axis to the power received by an imaginary isotropic antenna.

The gain of the dish antenna is

i. Directly proportional to the mouth of the parabolic reflector called antenna aperture. It can be calculated from the diameter (d) of the reflector and is equal to area of circle i.e., pd^2.
ii. Depends upon aperture efficiency which is a dimensionless parameter between 0 and 1.
iii. Inversely proportional to the square of wave length of microwaves.

Thus the larger is the aperture compared to wavelength, the higher is the gain of the antenna. Large parabolic antennas used in space craft communication and radio telescopes have very high gain aperture efficiency. The factors responsible for reducing the aperture efficiency are:

i. Radiation from the feed antenna falling outside the edge of the dish antenna called feed spill over.
ii. Intensity of the beam radiated by a parabolic antenna is maximum at the center of the dish and falls with distance from the axis.
iii. The feed antenna and its support block some of the beam and reduce the gain. It is called aperture blockage. This effect is reduced by using offset feed.
iv. The defects in the shape of the dish reduce efficiency. These are called shape errors.

Beam width

The angular separation between the points on the beam radiated by high gain antenna is defined by its angular width. The power drops to one half (-3dB) of its maximum value at the two edges of the radiated beam. For parabolic antennas, it depends upon the shape of the reflector and the feed illumination pattern. The relationship between gain and beam width is opposite. For a dish, radiation pattern of the feed antenna has a direct influence on the aperture efficiency and consequently on the antenna. It is accordingly tailored to the shape of dish.

The radiation from the horn which falls outside the edge of the dish is wasted and therefore, the gain is reduced. The side lobes are also reduced which cause interference in transmission and ground noise in receiving antennas. Most feed antennas are designed in such a way that the power radiated by the feed horn is 10 dB less at the edge of the dish antenna as compared to the maximum gain at the centre of the dish antenna.

Very narrow beams can be produced by parabolic antennas. For earth to satellite communication and back, very short aperture terminal (VSAT) are used. These are described in Chapter7.

Microwave receiver

Microwave receivers are called super heterodyne receivers having a low noise figure. The word heterodyne means mixing of two signals of different frequencies. Due to this a 'beat' frequency occurs as (i) the sum frequency or upper side band and (ii) difference frequency or lower side band. However, the upper side band is generally discarded in microwave receivers with the help of filters. A low noise figure is required so that weakest signal can be detected in the presence of noise generated within the receivers.

Microwave signal from the antenna is sent to the receiver. It consists of the following stages:

i. Filter to choose the frequency band as required
ii. Receiver protector (limiter)
   It protects the low noise amplifier (next stage) from any stray high power signal.
iii. Attenuator to reduce large signals which may cause saturation in low noise amplifier.
iv. Low noise amplifier increases the gain of received signal.
v. **Other parts of the receiver are as follows**

Mixer

Mixer creates both the sum and difference frequencies of the radio frequency and local oscillator signals. It is therefore used to create the intermediate frequency. It then converts them to lower frequencies by a filter. By this method, there is saving in the cost as lower frequency components are cheaper. Also filtering of unwanted signals at lower frequencies is easier due to higher band width. The unwanted frequency is called the image frequency. Example

Microwave frequency	12 GHz
Local oscillator frequency	10 GHz
Upper beat frequency	22 GHz
Lower beat frequency	2 GHz
Image frequency	8 GHz

This unwanted image frequency can be prevented by using an image filter.

Local oscillator

The local oscillator signal should be maintained at a fixed or minimum power. Sometimes an amplifier is required in the local oscillator circuit. This, however, needs to be kept in suppression mode to reduce variation in the oscillator power coming from the input.In order to reduce noise figure, wave guide components are used before the low noise amplifier.

Microwave circuit components

Just as transmission lines are used for transmission of electrical energy, wave guides are used to transmit microwaves. They are hollow conducting tubes generally rectangular in shape. The walls are perfect conductors and do not allow electromagnetic energy to escape. The wave travelling along a guide has a phase velocity and is attenuated. They are reflected when they reach the end of the wave guide. This can, however, be eliminated by the use of a impedance matching system. If there are irregularities in a wave guide, the waves will be reflected. If there are holes in the wave guide, electromagnetic energy will escape outside. This method is used for transmitting microwaves from magnetrons to other units. If we close one end of the waveguide by silver or highly reflecting metal, both incident and reflected waves can cause a standing wave pattern. Many different fixed configurations (electric as well as magnetic) can occur in a wave guide. The interesting property of a wave guide is that a particular mode can propagate through the wave guide without significant attenuation only if the wave length of microwaves is less than some critical value which depends upon the dimension of the wave guide. This is the reason that the wave guides for 10 cm wavelength are bigger in dimension than 3 cm and still smaller for millimeter wave microwave propagation. The cut off wave length in rectangular wave guide is twice the width of the guide. Thus to transmit 300MHz, the wave guide width should be greater than 75cm. Such frequencies therefore cannot be transmitted by wave guide as it would be bulky and also not economical.

It may be noted that wave guides are widely used for the transmission of power at wavelength of 10 cm or less. The power losses in a waveguide are much less than the coaxial lines which use dielectrics with supporting insulators.The concept of wave guide is extended for microwave propagation in the atmosphere where temperature and humidity changes in such a way that a wave guide may form (Chapter 2).

Wave guide attenuators

Wave guides operated at wave length greater than cut off are often used to attenuate signal in a signal generator. These may be circular or rectangular. It can be extended by a coil where axis is at right angles to the axis of the guide. The pickup system also uses a similar coil.

Cavity resonators

These are used at extremely high frequency and are analogous to coil and capacitance used in communications. They can take different forms because any enclosed surface forms a cavity resonator. The simplest is a rectangular or circular wave guide shorted at each end. Each cavity resonator has an infinite number of resonant frequencies associated with a particular cavity. The 'dominant mode' frequency is called fundamental frequency and higher resonant frequencies are called higher order modes. Cavity resonator can also be made from the coaxial lines.

Cavity resonators connectors

A cavity resonator can be connected to a wave guide or transmission line by a small loop to link magnetic flux lines in a desired more. Alternately a probe can be inserted. The electric flux induces a current in it.If a voltage is applied to the probe, it will produce electric fields inside the cavity which produce oscillations. Another method of coupling is by means of a hole or slot.

CHAPTER 4

Microwave remote sensing

Remote sensing is defined as collecting and interpreting information about a target or object without being in physical contact with it. The methods using electromagnetic energy such as light, heat and radio waves are used for this purpose. The remote sensing observations are taken from ground, aircrafts, satellites or clouds. The original form of remote sensing is the aerial photography which is still in use. The technologies were later extended to thermal infra-red and microwave bands.

Active and passive remote sensing

Remote sensing methods have been classified as

- Active remote sensing
- Passive remote sensing

Active remote sensing method includes sending microwaves at the object and then detect them after reflection or scattering from the target. Radar is a typical example of active method. On the other hand, passive remote sensing system record the energy naturally radiated or reflected from an object. Thus passive remote sensing methods detect microwaves which do not originate from the satellite or any other instrument unlike radar.

Microwave remote sensing methods have been applied to study land, ocean, atmosphere and space. The sensors used for passive remote sensing are mounted on a satellite, aircraft or any other platform which collect microwave radiation emitted by the objects. These instruments are called radiometers. The sensors are further clarified as imaging and non imaging type.

a. **Active microwave sensors**
    i. Non imaging radar
    ii. Scatterometer
    iii. Radar altimeter

b. **Imaging radar**
    i. Real aperture radar, including Doppler radar
    ii. Synthetic aperture radar
    iii. Side looking air borne radar

c. **Passive microwave sensors**

The radiometers can be scanning or non scanning type. These can provide temporal or spatial resolution.

Applications of remote sensing

Land based

i. Soil moisture
ii. Crop identification
iii. Snow and glacier studies
iv. Forestry
v. Hydrocarbon exploration
vi. Urban land use
vii. Geology including earthquakes, volcanoes, landslides.

Oceanography

i. State of sea, tide and swells, wave height
ii. Sea surface salinity
iii. Sea surface temperature
iv. Wind speed over sea surface
v. Ocean circulation
vi. Shallow sea topography

vii. Oil pollution monitoring
viii. Geophysical surveys
ix. Ocean geoids studies

Atmospheric

i. Profiling of temperature and moisture
ii. Minor constituents like aerosols
iii. Spatial view of rainfall
iv. Weather forecasting
iv. Climate change studies

Space Rcscarch

Structure of the upper atmosphere, surface processes over planets, meteors and inter planetary space

Active remote sensing instruments

(i) **Radar**

The word 'radar' is an abbreviation from the phrase, "**radio detection and ranging**'. The principle of radar is based on echo ranging system for sound waves. When we shout at the surface of water in a deep well, we hear echo of the same after some time due to reflection of sound waves from the bottom of the well. Similarly objects like clouds, ships, aircrafts, mountains and buildings reflect microwaves back to the transmitter from where they are sent. The time delay between the incident and reflected wave depends upon the distance of the target. Since microwaves travel in air with the speed of light ($3x10^8$ m/s), each micro second of delay is equal to a round trip over a distance of 149.09 m (492 ft). The direction of the reflecting object is also found from the orientation of transmitting and receiving antenna. A radar can be used on ground or flown in an aircraft or satellite.

Some applications of radar are given below:

i. Weather to locate clouds, rainfall or estimate wind speed.

ii. Air traffic control; detection and guidance of aircrafts for landing and takeoff.

iii. Height measurements of an aircraft above ground by radar altimeter.

iv. Navigation information on location and direction of vehicles, ships and aircraft.

v. Police – detection of speed of vehicles

vi. Mapping terrain and geographical mapping of the ground, ice mapping

vii. Space exploration – study of stars, planets and natural satellites.

viii. Military

   a. Army for artillery, mortar and mine detection.
   b. Controlling the flight of a ballistic missile to target
   c. Identification of enemy aircraft or missiles or artillery and their surveillance.
   d. Naval warfare, anti submarine and anti surface war fare and navigation.

ix. Search and rescue locating aircraft, ships and people in distress.

Design of a radar

A typical radar system consists of

a) A transmitter

   i. Microwave oscillator to generate periodic pulses of high power but of very short duration.
   ii. Parabolic antenna to send or direct microwaves towards a target.

b). Radar receiver is a super heterodyne receiver with lowest noise figure, high sensitivity and proper band width. It receives microwave signals from any target which have parabolic antenna.

c). Duplexer: Generally a single antenna is used for transmission and reception of microwave signals. For this purpose, a switching mechanism called duplexer is provided. It isolates the sensitive receiver from the damaging effects of high power transmitter pulses generated by magnetron. During the reception of the reflected waves from the object or target, the duplexer switches the antenna to the receiver.

d). Computer interface and graphics: The analogue signals are converted into digital forms and stored in the computer. A large number of computer graphics can be made to identify objects or clouds.

Mixers are used to transform signals in one spectrum range to another spectrum range. In microwave transmitters, mixers are used to transform intermediate frequency signals produced by the wave form generator into radio frequency signals. In radar receivers RF signals are down converted into IF signals. Mixing is accomplished by combining the RF or IF signal with another signal of known frequency from the local oscillator. The resolution of the radar image depends upon the beam width of the antenna and the pulse width of the radar. The received echo energy is inversely proportional to the fourth power of the range. Thus to increase the range twice, the peak power needs to be increased 16 times. On the other hand, a larger antenna has a larger capture area and therefore increases the range by concentrating the transmitted power in a narrow beam.

It may be noted that long pulses give greater range than short pulses of the same peak power. But short pulses are necessary for seeing close target. Also, a weak echo is more easily distinguished if the pulse repetition frequency is higher. Range of a radar primarily depends upon average power instead of peak power.

The effective target area depends on its dielectric properties and generally increases with the size of the target. The noise figure is important to determine the performance of a radar system and is achieved by improved type of receivers. Most radars operate in the frequency range of 100 to 25000 MHz.

(ii) **Doppler Radar**

Doppler's principle is applied for sound waves reflected from a moving object. This principle is also applied to microwaves reflected from moving objects like clouds, aircrafts or insects. The need for such radar arises due to the fact that it is difficult to distinguish a moving target due to the presence of permanent echoes from cities and mountainous regions (if any). Microwaves are therefore produced by a radar called continuous wave or Doppler radar. If the waves are sent towards cloud drops moving away from the antenna of radar, the frequency reflected by these drops is lower than the frequency transmitted. On the other hand, waves from an object moving toward the antenna change to a higher frequency. The computer in the Doppler radar uses the frequency change to show directions and speeds of winds blowing around the rain drops, insects, aircrafts and other objects which reflect radio waves.

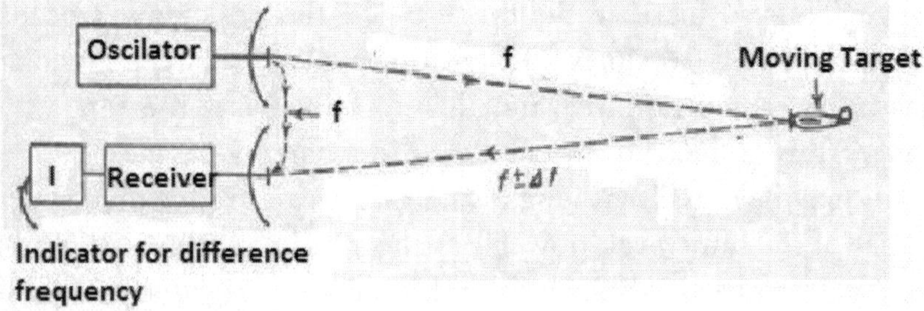

Fig. 4.1 Principle of Doppler radar

Radar images from clouds are colour coded according to the intensity of precipitation. The reflectivity depends on precipitation intensity as well as its type. For example, hail and sleet consisting of ice reflect more radio energy. Doppler radar is used for detecting a thunderstorm, tornado or cyclonic storm from the type of images they produce. A tornado has a vortex signature on radar screen while a cyclonic storm shows a spiral band of clouds. Thus, radar is capable of detecting wind shear in which wind directions change in small area and show rotation. The radar has also various other applications like aviation and military.

(iii) **Synthetic Aperture Radar (SAR)**

It works on the same principles as radar but it records the Doppler history as it is placed on a moving platform such as aircraft or satellites. It employs a small antenna that transmits a relatively broad beam. By recording the amplitude and phase of the microwave signal, the antenna is synthesized as if a single antenna would gather signals at different positions at different times. The signals are stored and then combined with specific phase shifts. This method gives a simulated (synthesized) set of signals which are equivalent to that received simultaneously by an actual large aperture phase array. To get further insight into its working, it may be noted that as the data is gathered originally with the distance travelled by the radar on moving platform, it varies with time and then gets converted into temporal frequencies. The distances that radar waves travel to and back from each scene element consists of some integer number of wavelengths plus some fraction of a wavelength. These fractions give rise to differences between the phases of the re radiation received at various SAR positions. The primary requirement of such detection is to find the differences between the phases of the received signals and the simultaneous phase of the transmitted signal. Since the track resolution improves, SAR gives better results as compared to real aperture radar (Fig 4.2).

In the Earth Resources satellite-I (ERS-I), the synthetic aperture radar gathers strips of high resolution imaginary 700km in width of the right of the satellite track. The 10 meter long antenna is aligned parallel with the satellite orbit to direct a narrow radar beam sideways and down wards onto the earth's surface. The imagery is built from the strength of the return signals which depends upon the roughness and dielectric properties of the surface.

SEASAT was designed to study mainly oceanic phenomena such as its roughness, current patterns and sea ice conditions using L wavelength.

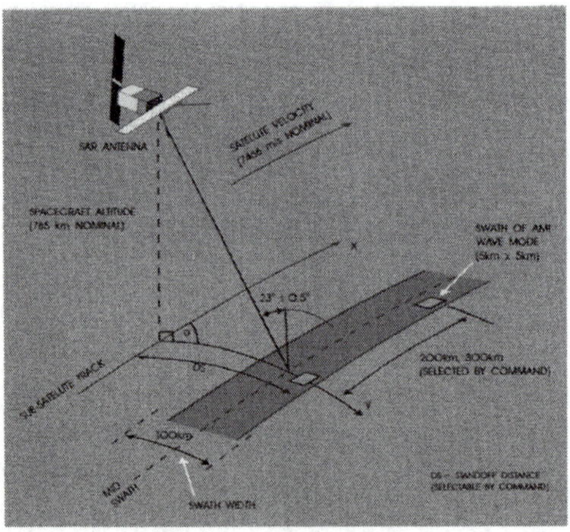

Fig. 4.2 Principle of SAR(From ERS-1, European space agency, France)

SAR interferometry (INSAR) is a modified version of synthetic aperture radar placed in satellites. In this technique, the phases of two SAR images of the same terrain are made to interfere to generate interference fringes. These two images can be taken simultaneously by a satellite with two antennas separated by a distance called base line. This is called single pass interferometry. Some satellites like ERS 1/2, do not have capabilities to acquire both the scenes simultaneously. In such a situation, two pictures of the same area are taken in different passes of the satellites.

In repeat pass interferometry, the base line varies from place to place and from pair to pair. However, optimum base lines values are taken depending upon the application like topography, surface change and surface movement . Appropriate pairs of SAR scenes are selected to apply the method. Similarly the time gap between two passes should not be large otherwise the data cannot be correlated properly. The images are obtained from the intensity and also with phase changes of waves.

The two images are used to produce interference fringes. The intensity and phase of two images are recorded accurately to $1/100^{th}$ of a pixel. The phases are subtracted from one another to get interference fringes. These are registered by different techniques. The fringes are filtered. These fringes due to flat terrain are disturbed by those due to topography (elevation) of the terrain. Therefore, after the flat terrain fringes are removed, we get the residual fringes which bring out topographical changes. Corrections are needed for the following :

i. As height increases, phase also increases. Since they are periodic and get wrapped after reaching angle of 2 pi, this is to be corrected to get unwrapped phase.
ii. Since radar measures slant range from satellite, instead of the true horizontal distance along the ground, this results in a varying image scale moving from near to far range. This

effect increases with the incident angle similar to increase of shadow as the sun sets.

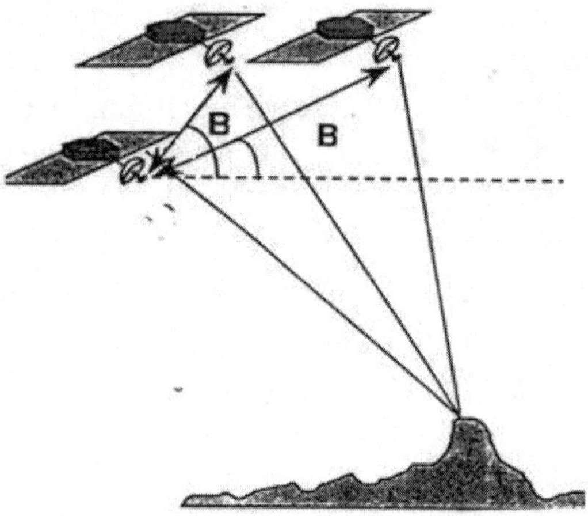

Fig. 4.3 Geometry of Differential IN-SAR.
B refers to base line of a pair.

In order to increase the accuracy of measurements, the concept of differential In- SAR is shown in Fig 4.3. In this case, three satellite passes are utilized to generate two interferograms. One interferogram is generated using passes 1 and 2. Another interferogram is generated using passes 2 and 3. The difference of these two interferograms will give rise to differential interferogram. During these passes, if there is no change in the elevation of terrain, the differential interferogram gives zero. On the other hand, if there is some change in land elevation due to earthquake or volcano movement, the phase values will become different. Using this, topography mapping may be done. Sometimes, only two In SAR scenes are used and the digital elevation model values are taken from other methods.

(iv) **Radar Altimeter**

It measures the altitude or the distance between the aircraft or space craft and the ground directly below it. It is therefore, different from barometric altimeter which measures the height of the aircraft above mean sea level. The principle of radar altimeter is exactly the same as used in a radar to measure the distance of the aircraft or a cloud (Fig 4.4). When higher accuracy is required, continuous wave radar using Doppler's principle is used.

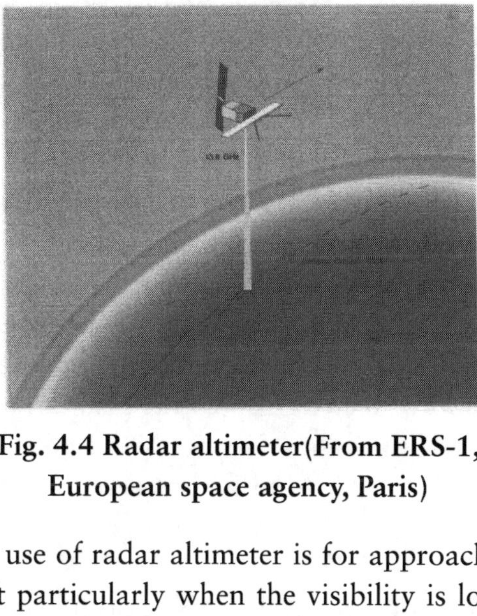

Fig. 4.4 Radar altimeter(From ERS-1, European space agency, Paris)

The main use of radar altimeter is for approach and landing of the aircraft particularly when the visibility is low due to fog or dust. It is therefore, used for automatic landing of the aircraft through a flight computer.

(v) **Ground penetrating radar**

Ground penetrating radar(GPR) is widely used to study shallow subsurface structures like faults(weak zones) or hidden objects. It provides high resolution mapping of the objects underneath ground surface. GPR uses antenna frequencies in the range of about 50 to 200 MHz. The antenna can be dragged on the ground surface by a vehicle or by a technician(hand towing). Microwaves are sent inside the ground by antenna in contact with the ground. They are reflected from a discontinuity from a material of different dielectric constant like junction of soil and rock, cavities, voids, filled area or any buried object. They are received by an antenna. The time difference is used to locate its depth like in other radars. The strength of the echo depends on the absorption of microwaves in the media between the radar and the discontinuity, shape and size of the target. Multiple observations along its path are used to create images of the structure below its surface. The depth of ground penetrating radar can be 10 to 15 meters in sandy or silty soil by using low frequency radar antennas. Higher frequencies enable to examine in greater detail but their depth of penetration is limited. GPR can be used upto a maximum depth of 30m. Maximum depth of penetration is found over ice upto several hundred meters. The GPR signal does not go deeper in clay or moist soils or salt contaminated due to their higher electrical conductivity and is considered a limitation of the method.

GPR observations along a line give a profile or cross sectional view of subsurface. The data collected from multiple lines collected over an area are used to construct three dimensional images which is also called tomography.

Applications of GPR

i. Geological studies including subsurface sedimentary structures, faults

ii. Archaeological studies – to find hidden remnants of forts, objects to study historical past and culture. Such studies were reported to have been undertaken at the Ayodhya site
iii. Landfills, burial trenches and grave yards
iv. Landmines
v. Underground pipes, tank and drum location
vi. Concrete inspection
vii. Any buried object

(vi) Radar Scatterometer

A radar scatterometer is a non imaging active system to measure radar backscatter as a function of the angle of incidence. A scatterometer transmits a continuous radar signal directly along the flight path. The beam illuminates both ahead and behind the aircraft or satellite. The zero angle incidence directly below is illuminated initially by the aircraft and 60 degree is generally the maximum angle of incidence. The signal returned depends on the backscatter properties of the terrain. As the aircraft or satellite advances, a ground resolution cell is illuminated initially at 60 and then at successively decreasing angles. The recorded amplitude and frequency of the successive returns are processed using Doppler method to obtain scattering coefficients at each of several incidence angles. The wavelength used in scatterometer is generallyKu band(14 MHz) for wind measurements.

Applications Radar Scatterometer

i. Wind measurements used for air sea interaction, climate studies, monitoring hurricanes or tropical cyclones (Chapter 10)
ii. Study of vegetation, soil moisture, polar ice, sand and snow dunes

iii. Non terrestrial applications like moons of Solar system using space probes like Cassini mission(Chapter 15)

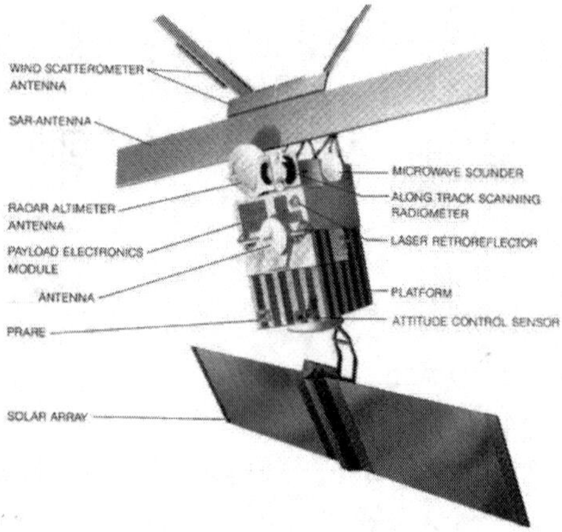

Fig. 4.5 Microwave measuring instruments in ERS -1

Microwave measuring instruments aboard satellite ERS 1 are shown in Fig 4.5.

Passive microwave imaging

Energy at microwave wavelengths is radiated from surfaces just like thermal infra red wavelengths.But the energy flux defined as rate of flow of electromagnetic radiation measured in watts per square centimeter is low at these long wavelengths (**Wien's Law**). This energy needs to be sufficiently large to be recorded which is done by taking large ground resolution cells of the order of kilometers. The radiant microwave energy is recorded by the scanner systems with detectors operating in the microwave region to produce images of microwave brightness temperature. This microwave temperature of a surface is the product of kinetic temperature (as measured by a thermometer) and emissivity(ratio of radiant flux from a body to that from a black body at the same kinetic temperature) at the wavelength band of the detector.

Such passive microwave imaging systems have been deployed on several environmental satellites.

Microwave radiometer

Microwave radiometer is used to measure energy emitted in the microwave region or brightness temperature of the target to measure atmospheric and terrestrial radiation, meteorological or oceanographic conditions.

The microwave radiometer broadly consists of

i. Antenna
ii. Low noise amplifier
iii. Radio frequency isolator
iv. Intermediate frequency oscillator and amplifier
v. Mixer to generate beat frequency
vi. Frequency amplifier and filter
vii. Square law detector to convert into direct current (D.C) and amplifier
viii. Integrator
ix. Computer interface

The scanning multichannel microwave radiometers mounted on satellites measure in five microwave wavelengths through its different channels. Its instrumentation consists of antennas and reflectors around a multi frequency horn. The power is concentrated into the feed horn which reflects multiple frequencies. The scanning is done at few km per second with a scan time in seconds. The data is sent from the satellite to earth data center and then processed, disseminated and archived.

Microwave radiometers measure temperature calibrated in degree Kelvin scale (273+^{0}C as absolute zero). The calibration procedure of radiometers is quite complex for which corrections need to be applied.

The following surface and atmospheric parameters can be computed from the energy emitted at that wavelength by microwave radiometers.

i. Air temperature
ii. Sea surface temperature
iii. Sea surface winds
iv. Salinity
v. Soil moisture
vi. Sea ice
vii. Precipitation (rainfall, snow)
viii. Total amount of water vapour.
ix. Total amount of liquid water directly below or above the instrument.
x. Cloud observations
xi. Low level wind shear

Ground based microwave radiometry can be used at two frequencies which allow simultaneous determination of water vapour and liquid water content along a selected path in the atmosphere. Water in the atmosphere emits microwaves which increase with frequency, dominating 31.4 GHz observations while water vapour dominates 23.4 GHz channel. Oxygen channels with frequencies in the range of 57 to 58 GHz have been shown useful to capture low level wind shear(below 500m). A ground based microwave radiometer with several channels gives many parameters listed above.

CHAPTER 5

Microwave applications in daily use

Microwaves have revolutionized our daily life through their application in ovens, television and cell phones. Their use is increasing day by day all over the world. Of these, microwave oven generates high power or radiation locally and can be installed anywhere.Television requires one way communication via satellite but can also be set up in most of the places by the use of dish antenna through which microwave signals carrying video and sound signals are received. However, cell phone requires both terrestrial as well as satellite communication net works. It is therefore, discussed later in Chapter 7.

Microwave oven

Microwave oven is a kitchen appliance used for heating or cooking food. It is less hazardous than gas or stove cooking and is safer even for grown up children if they are trained to operate it carefully. Several food items can be warmed simultaneously by placing them in separate containers on the rotating base called turn table (Fig 5.1).

Microwave ovens, however, do not directly produce browning effect which require very high temperatures. Higher temperatures required for frying, browning or baking are not achieved in the oven. Modifications have been introduced in microwave oven where different kinds of heat sources are installed inside to produce heating effects to produce browning.

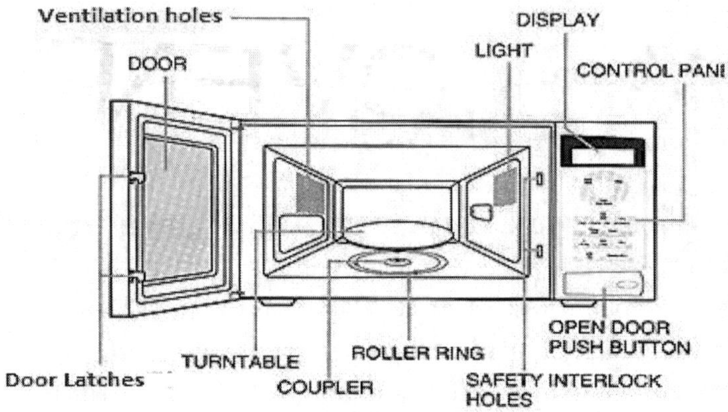

Fig. 5.1, Main parts of microwave oven (Samsung); electronics not shown

Mechanism of heating

Gas stove and kerosene heat a substance directly when ignited. Electricity has long been used in electric heaters to cook food. But heating effect inside microwave ovens is produced by high frequency microwaves of 2.45 Gigahertz (GHz) or wavelength of 12 centimeters generated by magnetron. It may be interesting to understand how such heating is produced inside oven. Most substances are made of polar molecules. In these substances, the positive and negative charges are at a distance from each other. An important property of these molecules is that they try to align themselves when alternating electric field of the microwaves is passed through them. In effect, the molecules tend to rotate and collide with other molecules. This absorption of energy from microwaves is manifested as molecular vibration in solids and liquids and is given out as heat. These substances are also called dielectric substances and the effect is described as dielectric heating. In simple words, microwaves force water and fat molecules in food to rotate. This forced rotation produces heating and used to cook food. It is important to remember that depth of penetration of microwaves is dependent on composition of food. The lower is the microwave frequency, the larger is the penetration effect. The microwave heating frequency of 915 MHz equally heats water as well as food. The type of food item and its quantity often requires some experience to adjust optimum heating. Different food items convert microwave radiation into heating to different extent. This selectivity allows some part of the substance to heat more quickly or more slowly than others.

Foods high in water content and with small quantity of oil rarely reach temperatures greater than the boiling water. However, some substances like sugar, fats and oils (triglycerides) attain much higher temperatures inside microwave ovens due to their lower heat capacity and higher vaporization temperature. In some cases, it can produce effects like browning or grilling. Microwaves can cause melting in ceramics and certain type of rocks and glass and therefore special utensils are made for microwave oven.

Design of microwave oven

A microwave oven consists of :

i. A power source of high voltage like a transformer or an electronic power which converts 220 volts electric current into high voltage to operate a magnetron. For this purpose, a high voltage capacitor and diode connected to the magnetron transformer increase supply voltage of 220V AC to about 3000 volts required to operate a magnetron.

ii. A cavity magnetron to convert high voltage electric energy to microwave radiation. This is located either at the back or in one side of the oven chamber.

iii. A micro controller (magnetron control circuit)
iv. A wave guide to direct the microwaves towards a rotating plate on which food is kept inside cooking chamber.

A visible light is seen inside chamber when microwave oven is switched on. When timer runs out, the oven turns off. The cooking chamber is of special design. It prevents microwaves from coming out of the oven. The window of oven door has a layer of conductive mesh to maintain shielding. Due to small size of perforations in the mesh which is much less than the wave length of microwaves, most of the microwave radiation do not pass through the door. The visible light, however allows us to see inside objects due to its very short wavelength.

Modern microwave ovens generally use a digital control panel for operation and use LED, liquid crystal or vacuum fluorescent display. Numeric buttons are provided for setting clock time, power level selection and pre programmed settings for different food items like meat, fish, frozen vegetable, pop corn. The extent of heating depends upon the electric power used and the type of material. For producing browning effect such as in cakes, the design of microwaves ovens is modified so that it is a combination of standard microwave oven and conventional electric oven. In such models, high power quartz halogen bulbs are fitted which emit infra red radiation to heat the surface of food items. The browning effect is produced due to infrared rays on reactions in food containing carbohydrates and proteins. For this purpose, a special tray made of glass or porcelain is often provided. Microwave pop corn bags are often packed in a thin aluminum foil or paper trays. The thin metal film efficiently absorbs microwave energy and consequently, becomes extremely hot. It radiates infra red which concentrates heating in oil used in popcorn. Such heating packages or trays should be thrown away as waste after using once.

Special cookware in microwave oven does not get directly heated due to the property of its material making it transparent to microwaves. This results in lesser burns as compared to conventional ovens which heat the cookware as well as food to the same extent. Also, microwaves penetrate deeper than direct heat in conventional ovens.

Ordinarily, a kitchen microwave oven consumes 1100 watts of electric power. Due to the dissipation of heat in the magnetron tube, only 700 watts of microwave power is produced giving an efficiency of 64%. Additional power is used to operate the lamps, AC power transformer, magnetron cooling fan, food turn table motor and control circuits. This heat generated along with that from the hot food is exhausted as warm air through cooling vents.

The smallest microwave ovens operate with 12 watts and can be fitted in cars to warm our food. Such portable 12 watts microwave oven includes a cool bag as mini refrigerator. Thermal liner keeps the items cool for hours. The cool bag should be removed from inside the microwave before operating portable microwave oven.

The use of metallic containers inside microwave oven is prohibited because they reduce microwave power to zero in a very short time. If we keep any object containing pointed metal in oven, an electric spark occurs when microwaves fall on it.

Rectification of faults (trouble shooting)

i. If microwave oven does not start, check power cord, plug and all connections.
ii. If on placing food, the turn table does not rotate; check the groove in the center and adjust its position inside holes.
iii. If turn table rotates but food is not get heated, the magnetron could be damaged. Its replacement may involve heavy expenditure.

Precautions to be taken while operating microwave oven

i. Never press 'start button' unless some substance is kept on the turntable. If turn table is allowed to run without any load to absorb microwaves, the incoming and reflected microwaves will produce standing waves. This may cause the magnetron tube to overload and burn out.

ii. If eggs, potato or tomato are to be boiled, pierce holes in them otherwise they may burst.

iii. If due to any reason, there is a spark or fire inside, never open the door. The fire will be extinguished in a short time due to limited availability of oxygen inside oven.

iv. Periodically clean turn table and oven walls of the microwave with moist cloth. It may be remembered that the turn table is properly mounted on its grooves and checked by giving it slight rotation.

v. Never keep any substance made of metal inside the oven.

vi. Use only 'microwave safe' plastic containers in the microwave oven.

vii. Try to reheat cooked food sufficiently to destroy bacteria which may grow during refrigeration.

viii. To warm chapattis, either wrap them in suitable plastic foil or place it directly on the cleaned turn table. However, not more than 15 to 20 seconds should be used for warming otherwise they become hard in a short time.

ix. If refrigerated rice is to be warmed, sprinkle a little water on its surface and then warm it inside oven. It will become tasty like freshly cooked.

x. Don't warm food inside utensils with tight covers otherwise they may be thrown out during heating. The locking arrangement in the plastic lids also becomes loose making it unsuitable for future use.

xi. By boiling a glass of water with lime juice in a microwave oven, the smell from the microwaves can be removed.
xii. Change microwave oven after a few years since in old oven, some leakage of microwave radiation may occur.
xiii. Follow cooking instructions as given by the manufacturers in the booklet.

In spite of extensive use of microwave ovens worldwide, a question has been raised whether food cooked or warmed in them becomes harmful to human beings. The heating by microwaves is caused by molecular vibrations of the substances. The heat produced may, therefore, cause internal damage to the molecular structure of vitamins and phyto nutrients. However, nutritional elements are rarely destroyed in microwave ovens. Minerals also mostly remain unaffected. Some people consider that food cooked in microwave ovens causes obesity but all such findings are not substantiated. Some hazards likely to be associated with microwave ovens, though controversial, are discussed in Chapter 16.

Digital television

Television uses a camera that successively explores or scans different portions of the scene to be reproduced. It produces a voltage proportional to light intensity of the portion of the scene scanned. The varying voltage that results from scanning is amplified and modulated upon a radio frequency carrier wave and then radiated from the TV station. The television receiver synthesizes the original scene on the fluorescent screen of a cathode ray tube. This is done by cathode ray spot to trace over successive portions of the reproduced scene according to scanning of the original scene. It also varies at the same time to vary the brightness of the spot in accordance with the envelope amplitude of the received signal The scanning process is done with sufficient rapidity which gives continuous picture.

Television using analogue signals has been in use for several decades. However, the problem of interference of analogue signals and lack of quality of pictures arise due to

i. Ghosting of images
ii. Noise for weak signals
ii. Degradation of the quality of image and sound.

On the other hand, digital broadcasts provide more digital channels in the same place.In digital television, the audio and video signals are digitally converted, combined together on the microwave carrier and then transmitted. Both these signals are synchronized digitally and therefore, the reception of the digital signal should be processed to make both audio and signal free from any disturbance. Several regions of the world are using different broadcasting standards.

Television pictures can have different pictures formats. It is dependent on the number of picture elements (called pixels) in each line for using different formats. Thus the scanning is done, say in format of 640x480 implying that there are 640 elements in each of 480 lines. The total pixels will be 307, 200 and width to height ratio called aspect ratio of 640 ÷ 480 or 4:3. High definition television that can be transmitted as digital terrestrial broadcasting can have 1280x720 pixels in progressive scan mode or 1920x1080 pixels in interlaced video mode. Each of these has aspect ratio of 16:9. The other range of format used for standard definition T.V can adopt several formats depending on the technology used in the country of broadcast. For the PAL countries such as India, 720x576 pixels are used for 16:9 broadcasts. However, 720x480 format is used in NTSC countries. For 4:3 aspect ratio, the formats will be 720x576 in PAL countries and 640x480 in NTSC countries. If these resolutions are reduced, there will be saving in band width. The rate of data broadcast may be 19 megabytes per second, but these can be sub divided . Instead of high definition television, there are advantages in using standard definition digital signal. In this, the band width of a multi layered channel is subdivided into several digital sub channels. This allows different television programs to be transmitted on the same channel.

Digital television can be viewed in the following ways:

i. Dish Antenna
ii. Digital cable antenna and digital satellite.
iii. Digital multimedia broadcasting allows mobile phones to receive TV signals
iv. Internet protocol TV (IPTV) uses internet and optical cable line
v. Receive digital TV via open internet. Internet television software can be used to watch TV on a computer.

Communication requirements for satellites sending programs directly to televisions in our homes is further discussed in Chapter 9.

CHAPTER 6

Microwaves in industry and environment

Microwaves have shown immense applications for the benefits of mankind. They are widely used in industry, security systems and environmental problems. Their new applications are being reported from time to time.

Microwave power

The use of microwaves for power transmission and solar powered satellites has generated interest in several countries. Based on a study in USA, it has been reported that 30 KW of microwave power can be transmitted over a distance of 1.6 km and converted to DC power with an efficiency of 84%.Microwaves may provide energy for future space transportation. The microwave light craft mission developed by NASA is powered by microwaves directed from an orbiting solar powered satellite. Magneto hydrodynamic lifting forces are used to lift the craft after breaking air molecules into a plasma. Microwaves are used in fusion reactors to heat the fuel into a plasma state using frequencies in the range of 110-170 GHz. Microwave heating is used in industrial processes for drying and curing products and in food preparations.

Pharmaceutical industries

The main advantage of microwaves as a source of energy for heating synthesis reactions has been demonstrated through several applications. The two approaches adopted are batch processing and continuous flow. In batch processing, the reagents are pumped into the microwave reactor. The reaction is run according to the predetermined parameters. The process is repeated to get desired amount of material. In the continuous flow systems, the reaction mixture is continuously pumped into the microwave cavity. However, the percentage of reactions being performed with microwaves as the energy source has been estimated as 10 to 25% only.

In the area of bioscience, the microwave instrumentation has been used in peptide synthesis. However, the use of microwaves in accelerating reactions and rapid heating requires additional safety measures. These include development of explosion proof reactors, shut down mechanism during over heating or over pressurization and venting mechanism for closed vessel reactions like pressure cooker.

Mineral processing

Mineral processing using microwaves has been envisaged. Their potential use lies in selective mineral liberation, leaching extraction like gold, iron, titanium and zinc and desulphurization of coal for cleaner combustion. New process using microwaves promises many commercial applications in mining industry.

Non destructive testing

Usage of microwaves for testing materials is based on the fact that dielectric constant and dielectric loss change in materials filled with different amounts of air, humidity or moisture. Thus by comparing dielectric properties of pure substances with those mixed with any impurity or air gap(void) or water, one can easily detect any impurity or defects in construction materials like concrete, mortar, cracks in metal surface or defects under paints. Australian scientists have developed a CSIRO SpeedGrader to grade timber using high speed scanning through microwaves and discard wood with knots and other defects. This method is cheaper than X ray technology.

Researches in chemical processes

Microwaves have different properties in materials like solvents, catalysts, reagents which also change with temperature due to their energy dissipation factors converting their energy into thermal energy or heat. They are used to accelerate chemical synthesis and reactions.Microwave absorption spectroscopy, magnetic resonance spectroscopy and electron spin resonance spectroscopy provide many applications in chemical research like organic synthesis.Microwaves are also used to generate plasma for reactive ion etching and plasma enhanced chemical vapour deposition.

Microwave security systems

For checking luggage, X rays are widely used at airports and railway stations but cannot be used to check human beings as they are harmful due to their ionizing nature. The human body is therefore screened using millimeter wave microwaves to detect suspicious metallic and non metallic hidden objects. Microwaves are also used to protect important places like military bases, government and industrial buildings and nuclear power plants through various security systems.

Microwaves of frequencies 800 to 950 MHz are also used for identification and tracking through radio frequency identification tags. They use combined radar and radio technology and consist of a transponder or a tag containing microchip and antenna. They can be placed on the object to be identified. The transmitted data from the tag is then read. Its applications include fare payment on toll roads, parking in addition to security and identification cards. Bar code scanners can track shipments or transportation by road.

High power radiated microwave energy can damage microprocessors in modern cars which control central engine functions like fuel injection, fuel pump control and others. Such a device can be used to force vehicles to stop by traffic police or other law enforcing agencies and also for protection near military bases, communication systems and oil platforms.

For providing security, the use of CCTV cameras is increasing in our houses, offices, public places, shops, industrial establishments and other places. With Wi Fi through internet, we can monitor CCTV at any time in our house or office. The live recording by CCTV can also be monitored through mobile telephones outside our homes.

Environmental monitoring

Oceans

Pollution monitoring

Oil carrying ships sometimes collide with other causing spilling over ocean surface. The leakage of oil may also occur in oceanic oil platforms. This has adverse effect on sea fishes and other species. Oil slicks have a dampening effect on water waves. They can be detected by radar due to the difference in scattering; being low in oil film as compared to stronger back scatter from rough clean water. Satellites like ERS-1 supplement air borne radar systems as they can monitor large area of oil pollution.

Coastal processes

The behavior of the sea in proximity of coasts, the sea bed profiles and the changes in the coastlines (if any) are monitored by a combination of microwave instruments like radar altimeters, and wind scatterometers mounted on satellites. This information is obtained from all the places which are not easily accessible or covered with clouds.

Ship surveillance

The synthetic aperture radar helps in ship surveillance over extensive sea areas. The information along with GPS could be used by concerned countries to track hi jacked ships as reported for Somali pirates. Urgent measures can be taken soon after detection.

Sea salinity

The Aquarius satellite is studying the surface salinity of seas to provide a global view of areas covered by salinity to understand key climatic processes.

Forest monitoring

Satellite is being used to monitor the impact of deforestation and restoration methods of areas covered by forests.

Flood and drought monitoring

Satellite and other air borne techniques mounted with radars and other visible and infra red sensors enable us to monitor the areas of floods caused by heavy rain, damage to dams and also areas fulfilling drought conditions. It may however be noted that if thick strata of volcanic ash released from volcanoes persists in the regions, their results become questionable as reported during Mt. Pinatubo volcanic eruption in 1991.

Ice mapping

The day and night capability of active microwave instruments is particularly useful to map the distribution and motion of sea ice, mapping of the extent of glacial ice and surface features. These are valuable in climate applications besides marine navigation.

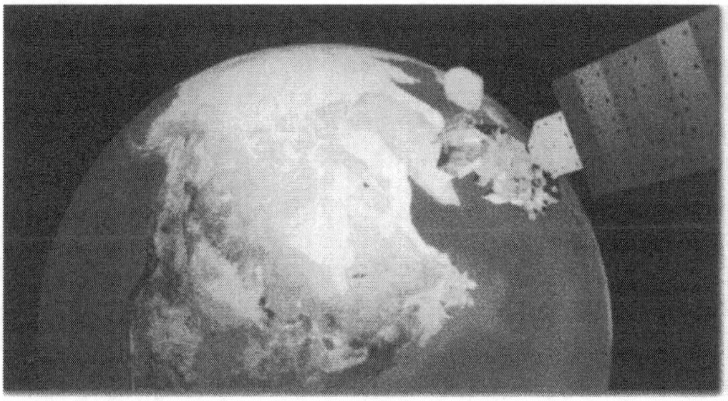

Fig. 6.1 Sea ice in polar region (ERS-1)

Ozone layer monitoring

Depletion of ozone layer in the stratosphere has been causing concern worldwide due its harmful effects on human beings and may cause skin cancer. This is attributed to the use of chlorofluorocarbons in refrigeration industry. The thickness of ozone layer is monitored worldwide by using millimeter waves namely very short microwaves.

The applications of microwave techniques in solid earth studies, and atmosphere have been discussed later.

Agriculture (viticulture)

While soil moisture and other parameters using microwaves have been discussed in Chapter 9, new application within precision agriculture has been evolved called precision viticulture. For this purpose, large homogenous fields are divided into smaller units based on field or other parameters which may need different methods of management. Its application has been shown in recent years to produce wine from grapes. The technology assists in research on soil nutrients variability, mapping and management, weed control, economic and environmental sustainability. Some of the technologies adopted are as follows:

i) Satellite / Arial survey for field estimation, special imagery for higher indices like leaf area index.
ii) Ground penetrating radar for soil water content

GPS and GIS enable grape growers to locate specific spots in their vineyard during sampling for vine growth, development and productivity, quality of vines, destroyed vines or rabbit holes.

Archaeological application

Archaeological investigations enable us to know ancient and extinct cultures through buried objects, tools and monuments. For this purpose, archaeologists carry out excavations on pre historic and historic sites. For archaeological search on regional, local and object scale, the techniques used are as follows:

i. High resolution satellite image
ii. GPR and lidar
iii. GPS
iv. Aerial, terrestrial images.

Disaster management

Satellites provide reliable and rapid communicable tool when ground infrastructure is damaged during natural disasters like earthquakes and cyclonic storms.Earth observing satellites send images or maps of disaster effected areas. GNSS provides positional information on disaster related events to support relief measures in the field. Satellite communication serves to warn people at risk in remote areas. It connects a disaster area to the outside world for tele medicine or other requirements. It does not fail as land communication. V SAT can be quickly installed in the disaster effected areas to get data, voice and photos. Portable base towers can provide rapid means of communication for this purpose.

Several international projects have also been formulated like SEAGAL project – south east Asia Centre on European GNSS for international cooperation and local development. It provides new link between Europe and south east Asia in the framework of satellite navigation and related activities like education and technology transfer. The research center is located in Hanoi, Vietnam. The United Nations General Assembly established UN-SPIDER Program implemented by the United Nations office for outer space to develop the capacity to use all types of space based information to support the full disaster management efforts. The lessons learnt from Haiti earthquake (January 2010) further improved the effectiveness of the program.

Research in new areas.

The advantage of using microwaves in laboratory to study dielectric properties namely dipole moment and relaxation times have shown new applications. The low loss dielectric material used in coaxial cable were found by this method long time back. The dielectric properties of rocks brought from moon have thrown light on their composition at different microwave frequencies. The samples when brought from Mars or outer space objects would reveal their composition for planning future studies.

CHAPTER 7

Microwaves application in medical science

Discovery of different types of waves by physicists led to their application in medical science ranging from diagnosis to treatment of human diseases. These waves are called infra red, ultrasound, lasers, X-rays, gamma rays and microwaves. Their application in medical instruments has been of immense benefit to mankind in the detection of abnormal symptoms ranging from fractures, gall bladder (stones), kidney, prostrate, liver, cancer and other diseases. Computerized tomography using sound waves, X-rays and radio waves generate a three dimensional image of the diseased organ providing better diagnosis. Two properties of microwaves are used in medical science. The first application is based on using microwaves as a source of power in medical equipment of radiology called diagnostic imaging method.The second is to use it as therapy due to its ability to create intense heat somewhat similar to microwave ovens used for cooking food.

Diagnostic methods

The ability of the ultrasound, X-rays and radio waves to enter our bodies led to development of imaging techniques. Their relative advantages or limitations are described below.

Ultrasound

Ultrasound is an instrument which sends sound waves inside the body and after reflection produces the image of the affected part. The different echoes thus obtained from tissues allow a three dimensional picture. For this purpose, a gel is applied on the patients skin so that the transducer or the probe can make better contact for the sound waves to pass through the body and give stronger signals for recording. This method enables to produce a shadow picture of the inner organ which needs to be examined. This technique is also used to assess the development of fetus during pregnancy. The method saves lives of women by early detection when pregnancy occurs in one of the fallopian tube requiring immediate surgery.It also helps in the detection of stones in kidney, gall bladder and the condition of other organs like liver, pancreas, kidney, bladder, prostate and testicles. The method has however, some limitation for fatty patient or if bowel gas is present between the probe and the organ. Ultrasound cannot penetrate bones. Before undergoing ultrasound, the patients are advised to drink lot of water and retain it without urination till the test is complete.This enables to obtain clear visualization of the organ. An enhanced technique called Doppler technology is sometimes used for ultrasound to measure blood flow in organs like liver or heart.

Computerised tomography (CAT scan or CT scan)

Xrays have been used since a long time to detect fractures and lung diseases.These are now used in a more sophisticated machine with the help of a computer to give cross sectional picture of the scanned organ. The method can identify normal and abnormal features of the organ. Sometimes contrast producing material containing iodine is used for CT scanning. Precautions are taken for persons allergic to iodine.

Magnetic resonance imaging (MRI)

Magnetic resonance imaging (MRI) gives different type of information about various parts of organs in the body than can be detected by X-rays, ultrasound or computer tomography scan (CT scan).The instrument produces cross-sectional images of organs and internal structures in the body.MRI is a medical imaging procedure which uses magnetic fields and radio waves. In most of the instruments, an electric current is passed through coiled wires which produces a temporary magnetic field around patient's body. Radio waves are sent by a transmitter and recovered by a receiver in the machine. The radio frequency current is briefly turned on producing a varying electromagnetic field of appropriate frequency called the resonance frequency. The magnet creates a strong magnetic field that aligns the protons of hydrogen atoms contained in water present in our body. Since they are also exposed to a beam of radio waves, the protons spin like a toy top. Their vibrations produce a faint signal that is detected by the receiver of the scanner. The output signal from MRI machine depends on water contents and magnetic properties of a particular area of the body which can be seen in the image. The frequencies used in MRI vary from 1 to about 170 MHz.

MRI scanner is a large cylinder containing a strong permanent magnet (field strength 1.5 Tesla) with a couch or bed that can move forwards and backwards inside. The pictures are taken inside the cylinder. The equipment is kept in a protected room to keep away radio waves which may interfere with the scan. Scanning is done with the patient lying on the bed who should breathe normally. Since MRI scans are very noisy, some protection for hearing is given. The radiographer monitors the scanning from the window of a neighboring room or on a T.V. screen. It may take about 30 minutes to one hour for detailed study during scanning.

Before scanning, sometimes contrast agent like gadolinium is injected. This helps to see abnormalities in the body organs more clearly. Precautions are taken for the people who may have allergy with the agent. MRI can be used on most areas of the body. Breast cancer may occur in some women. Its early detection has helped many women through timely treatment. The most popular method to screen breast is called mammography. It has, however, some limitations. A more advanced method uses microwaves which is based on the difference in dielectric constant between malignant and healthy tissues. The affected part appears to be inhomogeneous due to changing dielectric properties of the breast tissues. It can show soft tissues very clearly.This method has less adverse effect on the effected part as microwave radiations are non-ionizing. MRI is generally not taken for pregnant women. A single scan produces many pictures from different angles all round the body. It produces clearer results than a CT scan, X-ray or ultrasound to detect a tumor, diseases of liver and other stomach problems, fibroids, uterine abnormalities in women.MRI can also detect smaller tumors as compared to other techniques. MRI gives accurate information about the structure of the joints, soft tissues and bones of the body. MRI scan can evaluate the structure of the heart and aorta. Neurosurgeons use MRI scan for detection of tumor in the brain and other problems like internal bleeding and swelling or strokes. It is also used for assessing the structure of the spinal cord, vertebrae or inter vertebral disease of the spine. After head injury the patient needs to be immediately sent for CT scan and MRI. The Neurosurgeon then decides about the surgery of the specific spot in the brain. Fig 6.1(a) and (b) show brain tumors using CT scan and MRI respectively.

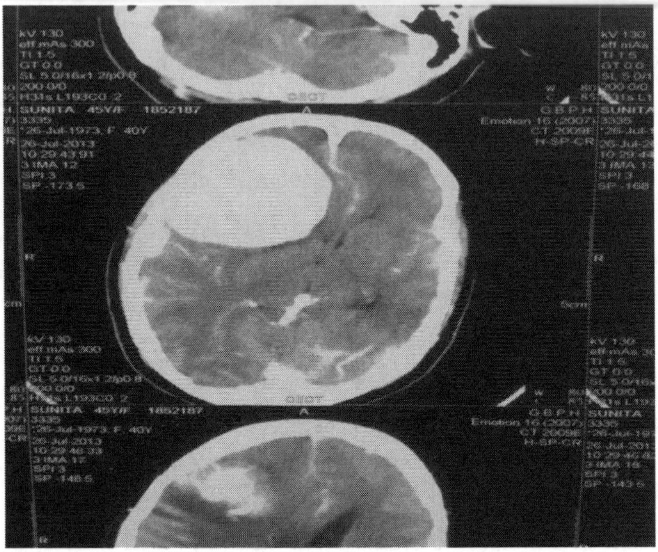

Fig 7.1(a) Contrast CT scan of brain showing large tumour in frontal lobe (Courtesy, Late Prof S.Sinha, Neurosurgeon)

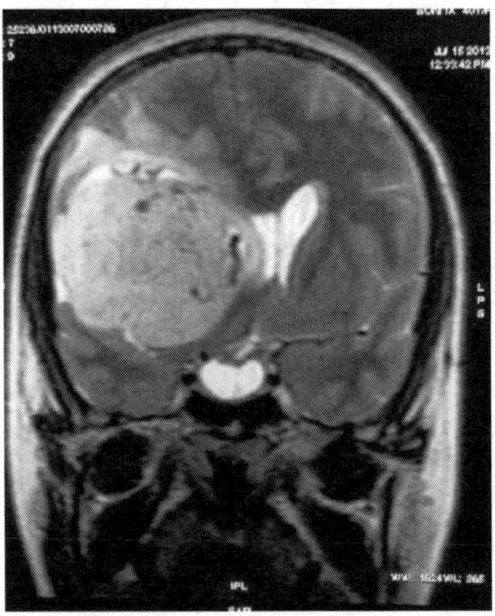

Fig 7.1(b) MRI brain T2 weighted(coronal cut) showing large tumour in frontal lobe (Courtesy Prof.S.Sinha)

Microwave therapy

Before the advent of microwaves, laser therapy has been widely used. This method uses intense beam of light to precisely cut, burn or destroy tissue. The laser beam is so small and safe that it poses no risks to the patient or medical team. It allows doctors to safely treat tissues without injuring the surrounding area. Lasers are used for the following procedures.

i. Surgery of prostate
ii. Removal of kidney stones
iii. Laser surgery of the skin
iv. Treatment of varicose veins
v. Corneal eye surgery to improve vision
vi. Repair of a retinal detachment

Treatment of cancer

Microwaves are used as a power source to accelerate electrons to high energy in an instrument called medical linear accelerator. The electrons are directed to collide a metal target. Due to the impact, the electrons are displaced from the atomic shells of the target producing X-rays. The high energy of X-rays are used to destroy cancer cells in the patient.

The heating effect of microwaves is also used for cancer treatment. This is called hyperthermia treatment in which the temperature of the cancerous tissue is increased sufficiently to kill it. In this method, the body tissue is exposed to a high temperature upto 44^0C. It is known that cancerous regions are usually characterized by regions of low energy content in tissues and low PH. Due to sufficient blood flow, the cancer cells become particularly sensitive to temperature between 40^0C to 44^0C. The damaged cells in the tumor are heated more easily due to reduced blood flow as compared to normal or healthy tissues. Thus heating by microwaves destroys cancer cells which is attributed to protein denaturation at temperature greater than 40^{0}C caused by changes in the molecular structure of cells and enzymes complexes. This method causes minimum injury to the surrounding healthy tissue. Radio therapy or chemotherapy is frequently used in addition to hyperthermia treatment.

Heart ailments

Microwaves are used to treat the condition of heart for irregular and rapid beating of the upper two arteries. This is called arterial fibrillation. In order to treat this condition, a small incision is made on the patient's chest. A special microwave sensor is placed on the heart. The surgeon treats irregular beating to restore the heart beat to normal. It is a much safer and faster method of treating arterial fibrillation which can be done in a few hours.

Trans and urethral microwave thermo therapy (TUMT)

Lower urinary tract symptoms are sometimes caused by benign prostate or benign cancer. The treatment is done by inserting a special microwave urinary catheter. Its heating effect destroys the surrounding prostate tissue after the procedure. Sometimes, the prostate becomes swollen and irritated. A catheter is placed to allow flow of urine without retention problem. After a few days (3 to 5), the catheter is replaced by a temporary prostate stunt to improve voiding which also reduces irritation symptoms. The procedure is done by urologist in about 30 minutes to one hour under the care of an anesthetic. Its side effects like urinary retention, infection and post procedurals pain need to be taken care of. The patient is discharged from hospital in a few hours after the procedure. He generally recovers in 4 to 5 days. Sometimes prostate oedema occurs after microwave therapy causing urinary retention. Then he needs to be kept in hospital for period upto 2 weeks or a temporary prostate stunt is placed after the first week. The stunt can be worn for 30 days. But urinary flow improves after a few months. Alpha blockers medicines are prescribed after TUMT which decreases incidence of retention and other urinary symptoms.Microwave therapy is an alternative treatment to trans urethral needle ablation of the prostates, prostate removal or prostatectomy.

Dementia

People suffering from dementia often do not remember their houses when they go out. Cases have been reported when such patients are found dead after a few days due to hunger, thirst and dehydration. It has therefore, been suggested that such patents should bc provided with global positioning system (GPS) so that they can be traced. Their movements from the house can be monitored further by infrared buzzer at the exit of door. A medical alert bracelet or a card with care takers phone numbers also helps in some cases.

CHAPTER 8

Microwaves application in communication

Microwaves have revolutionized television, telephone, mobile phones, radio and digital data handled by the computers. They are not only used for radio relay networks (terrestrial network) but by communication satellites. The high frequency of microwaves allows greater band width due to which more information of different types can be transmitted within the band. For example, the use of 10GHz microwave signal with a 10% band width provides a band width of 1 GHz. Thus radio signals (AM and FM), short wave radio, telephone calls, television and other digital data can be transmitted simultaneously. The short wave length of microwaves allows high gain antenna of smaller size to be used with narrow band width such as in a radar. The only limitation arises because microwaves require line of sight propagation. They cannot pass through mountains and other such obstacles. But these limitations are overcome by special networks.

Microwave transmission is used in:

i. Point to point communication system on the surface of the earth. It is widely used for mobile phones, television and data.
ii. Satellite communication
iii. Deep space radio communication

Different parts of the microwave band are used in radars, radio navigation systems and radar astronomy.

Microwave radio relay

This is also called terrestrial microwave communication. This technology is used for transmitting digital and analog signals such as long distance telephone calls, television programs and computer data between two places which lie on a line of sight path. The distance is of the order of 50 to 60 km between two sites. In microwave radio relay, directional antennas are used to get economical use of radio spectrum for radio connection between two locations.

Typical antenna used in radio relay installations are:

i. Parabolic antennas or dish antennas
ii. Dielectric lens.
iii. Horn reflector antennas

Microwave network design

This requires site acquisition, microwave network design, and interference analysis. Due to the requirement of microwave frequencies which travel in straight lines, the following aspects are included in planning a microwave network.

i. Line of sight connection between the two stations
ii. The first Fresnel zone free from obstacles to allow radio waves across a relay in uninterrupted path .
iii. High mountain peaks should be preferred for the microwave tower.
iv. Water bodies like lake or river in the mid path region are to be avoided due to distortion produced by two signals -one direct and the other due to reflection of waves.
v. Correction for multipath fading.

vi. The anomalies in the atmospheric pressure, temperature and humidity under different weather situation may cause large deviations and effect the transmission.
vii. High intensity rain and snow affect the transmission quality particularly if the microwave frequency exceeds 10GHz.

These factors called the transmission loss or path loss are to be taken into account while calculating their effects. Distances greatly in excess of line of sight range can be obtained by relaying. The signal over a path less than line of sight is re-transmitted over a second line of path and so on. These are also called repeaters (microwave towers). In order to maintain the link operative for a high percentage of time, suitable power margins are provided to main high standards in communication.

Fig. 8.1 Microwave towers

For microwave link to an island, the stray signal transmitted called 'toposcatter' is used instead of line of sight transmission path. The quality of signal in this method depends upon weather conditions and several other factors.

Transmission loss is measured in decibels. The decibel (db) is a logarithmic unit used in communication work to express power ratio. If P_1 and P_2 are power ratios

$$db = 10 \log_{10} \frac{P_2}{P_1}$$

Thus 1 db represents a power ratio of approximately 5:4 while 60 db represents a ratio of 1, 000, 000:1.On the other hand, if decibel is used to express amplification as in an amplifier, the power is proportional to the square of the voltage, E. Thus

$$db = 20 \log 10 \frac{E_2}{E_1}$$

The advantage of using decibels is that we do not need to remember large numbers because it gets reduced due to logarithmic scale.

Common network architecture

i. **Chain type.**
   For N stations, N-1 lines are required. Nth station depends upon N-1 lines.

ii. **Star type.**
   For N-station, N-1 lines are required. Each station depends on only one link.

iii. **Loop type.**
   For N stations, N lines are required. Route diversity is available for all stations. This arrangement is effective to maintain microwave communication during various faults which may arise due to
   a. Power failure
   b. Equipment failure
   c. Unexpected cut of cable
   d. Human mistake
   e. Rain and multipath fading.

Microwave links as discussed above are also used for communication between base stations and satellites. They are the back bone for television, mobile phones, short range indoor communications and mobile units fitted with camera during sports or even indoor recording.

VSAT

A very small aperture terminal (VSAT) is a two way satellite ground station with a dish antenna ranging from about 1 to 3 meters. (typically 0.9 to 1.9 meters). Satellites in geostationary orbit are accessed by them to relay data from remote earth stations called terminals to other terminals or master earth station called hub.

The hub consists of three elements namely radio frequency terminals, VSAT hub base equipment and the user interface. The hub enables the connection via satellite of all the VSAT terminals within the network and other communication from host to the VSAT.It is connected to the host computer. The other terminals function in mesh topology while the earth station works in star topology. Sometimes mixed topology is used to transfer data in the range of 56 Kbits upto 5 Mbit/s. They are used for

i. Sending broad band data to access remote locations with internet television programs
ii. Transmitting narrow band data for transactions from a credit card or polling data.
iii. Transportable or mobile maritime communications like television over a ship and other functions.

Now a days, Ku band VSAT are widely used for data transmission or telephone operations. VSAT consists of the following parts:-

i. Antenna
ii. Block up converter converts a band of frequencies from a lower to a higher frequency.

iii. Low noise block down converter converts a band of frequencies from a higher to lower frequency
iv. Ortho mode transducer splits signal between block up converter and low noise down converter
v. Interfacility link cable
vi. Indoor unit is a modem usually with ethernet port or special connectors to (a) transmit and (b) receive.
All the outdoor parts on the dish antenna are included in the outdoor unit.

The VSAT has two segments called earth segments and space segment. The later provides links to and from the satellite. This segment serves as a radio frequency repeater. The information(signal) from VSAT to the satellite is amplified and then retransmitted on a higher frequency. The hub in the VSAT network is essentially a network management system. It gathers data on the system, provides the system health check and other information like billing. The route terminal consists of two units. One unit is placed outdoor directly connected to the satellite while the other is placed indoors and connected to the user devices like computer.

Transmission loss

i. The transmission loss per unit length in wave guide or coaxial cable is independent of distance. It however, increases with frequency of the signal.
ii. The transmission loss in open space such as microwave links decreases with increasing distance. The transmission loss is expressed as the ratio of power transmitted or radiated to the power received. It is expressed in decibels (db).

For air to ground vehicular communication or navigation, both antenna have omni directional coverage meaning uniform coverage in the azimuthal plane. A monopole or vertical dipole is used for this purpose. For transmission between a ground tower and a satellite, the antenna in the satellite behaves an isotropic radiator. This takes care of spinning effect of satellite. For ordinary microwave links both antenna are made directional and decided by cost considerations. The received power is directly proportional to the square of the frequency.

These are ideal cases for free space transmission. But in practice, allowances need to be made for the effects of the ground, the troposphere and the ionosphere as described in Chapter 2. This is taken into consideration in calculating signal to noise ratio. Before planning microwave networks many field strength measurements at different microwave frequencies over various terrains like forest, hills, land and water were undertaken. Rain attenuation characteristics were also studied at microwave frequencies in India and other places in the world. These were also correlated with refractive index values computed from meteorological data which is readily available at most of the places (Chapter 2).

The sound, video and data can be transmitted by mixing them with a high frequency carrier by a process called modulation. The process involves operation of one of the three characteristics of the analogue carrier signal namely amplitude, frequency or phase. In a pulse system of radio communication, the pulses from the sampling process called digitization are modulated upon a microwave frequency carrier and then transmitted in the same manner as any other radio signal. These are then received by microwave antenna and a super heterodyne receiver with a band width corresponding to the pulses involved. The process of obtaining the original signal wave from the received pulses depends upon the type of modulation. In brief, digital modulation is transmission of digitally modulated analogue carrier signals in the communication. This method gives faster signal processing and has lesser effect of noise.

One category uses a constant amplitude carrier and carries the information in phase or frequency variations called PSK or FSK. Most devices utilize FSK modulation methodology. Another category conveys the information in carrier amplitude variations and is called ASK. These are used where multichannel voice communication is needed. They are used in military where mobility is required or in difficult terrain.

Satellite communication

Satellites have been launched since 1957 when the first artificial satellite was sent by the then Soviet Union. More than 50 countries have since participated to launch thousands of satellites which have a life span of a few years to decades. These are used for

i. Military purposes
ii. Earth observation
iii. Weather
iv. Communications
v. Navigation
vi. Data
vii. Research

Space stations and human spacecrafts in orbit are also called satellites to study planets and natural satellites like moon and titan. Each country assigns a name to the satellite launched like SPUTNIK (Russian), Apollo (USA), ERS (European), INSAT (India).

Geostationary satellites revolve around the earth at the same angular velocity as the earth at a height of about 35, 800 kilometers. They, therefore make one such revolution of the earth in 24 hours. Such a satellite appears to be fixed in one position to a ground based observer. Its advantage is that a ground based antenna directed towards the satellite does not require expensive system to track the satellite motion. Except in extreme northerly or southerly areas of polar regions, it provides communication applications to the rest of the world. They are positioned above the equator at a fixed place (i.e., longitude).

Another type of satellite called low earth orbiting satellite is placed in a circular orbit about 400km above the surface of the earth. Several such satellites are needed to cover large areas since only an area of about 1000km radius is visible from each satellite. Such satellites take only about 90 minutes to revolve around the earth. These are less expensive than geostationary satellites since they require low signal strength due to their proximity to earth. In order to resolve the communication problems in polar region, Russia designed a satellite called Molniya for the northern latitudes. It has a period of half a day.

Microwaves are used in the communication satellites. They use C, X or Ku band to communicate with the earth station. These contain following instruments:

i. Transponders, antenna and switching system.
ii. Power supply based on solar cells.
iii. Tracking sub system to a position of the satellite in desired orbit along with other adjustments like antenna direction.
iv. Communication sub system with ground control stations. Earth station receives the signals from the satellite. They also check its performance and functionality.

Different band width is needed for transmission for each service like TV, voice, internet, radio. The number of transponders is designed according to the requirements.

Television

Now a days, television programs can be recorded anywhere in our houses or open places. The dish antenna can be placed on a vehicle. The interview or any activity is recorded by video camera which is connected by a coaxial cable to the mobile antenna and relayed to TV station. This has made possible live telecasts of television news, sports, awards ceremony simultaneously on different channels by direct transmission of television signals in the microwave frequency band through satellite.These are then received in our houses through dish antenna. Mobile antennas are also mounted by a few air lines which enable the passengers to view TV programs on LCD screens mounted in the back of seats during flying or a large screen at various places in the plane.

Satellite radio

A satellite radio is a digital radio signal that is broadcast by a communications satellite. It covers a very wide geographical range than terrestrial radio signals.

The line of sight condition is necessary for the antenna to view satellites. If there are obstructions like tall buildings or parking places, repeaters are required to make the signal available to listeners of radio programs. The service is also available with roaming facilities which allows them to listen the radio programs anywhere. Satellite radio services are more expensive than terrestrial broadcast. They recover expenses through commercial advertisements but are free to the listeners of radio programs.

Amateur radio

Special satellites generally in low earth orbits have been launched to carry radio signals. These are accessed by amateurs with UHF or VHF radio equipment with dish or yagi antennas. The radio regulations of the International Telecommunication Union allow amateur radio operations in the frequency range 10.000 to 10.500 GHz and amateur satellite operation in the range of 10.450 to 10.500 GHz. Amateur radio has been found very useful in areas affected by natural disasters when other means of communication fails.

Satellite telephone

Satellite telephones are connected to either geostationary or low earth orbit satellites. Calls are relayed from land line telephones to an earth station. They are then transmitted to the communication satellite which is called uplink. From the satellites they are again transmitted to the earth station and hence to the public switched telephone network.

Satellite internet

Internet services via broad band data connections are used widely with laptop or desk top computers. The facility through communication satellites is not only useful at any place including cars, aircrafts, ships but also in remote areas in the country.

Local area network (LAN)

The major wireless LAN standard is called Wi Fi. Wireless base stations are connected to an ethernet network. These are able to transmit information over distances of several kilometers penetrating walls and non metallic barriers. This allows a laptop or desk top computer to connect to the Internet with a Wi-Fi adapter at any place in a house or office. LAN transmits information using 2.4 GHz and 5.6 GHz frequency bands.

A wireless router allows multiple users with Wi Fi to share a single internet connection. However, Wi-Fi routers, microwave ovens, cordless phones and blue tooth devices operate in the same 2.4 GHz frequency band. This may become congested and reduce the routers signal strength. It is therefore, preferable to use dual band router. Such a router will allow to use additional 5 GHz band simultaneously. Dual band smart phones and laptops can connect on 5 GHz while the rest may operate over 2.4 GHz.

Bluetooth technology

This technology uses 2.4 GHz microwave frequency for transmission of information for short distances. However, line of sight condition is necessary. It is extensively used in mobile phones, laptops and as an interface for wireless controllers in a play station.

Cell phones (Mobile phones)

Mobile phones are being extensively throughout the world which make our day to day life easy. Its users vary from domestic workers to richest. It makes and receives telephone calls over a radio link within a very large area. It is connected to a cellular network provided by a mobile phone operator.

The first commercial automated cellular network was launched in Japan in 1979, initially in Tokyo. The first 1G network was launched in the USA in 1983 by Motorola Company. Thereafter, it was extended to several other countries like, U.K, Mexico, Canada and then worldwide.

Mobile phones can be tracked by service provider and law enforcement authorities. On switching the mobile phone, the geographical location of a mobile phone can be determined using a technique called multi alteration to calculate the differences in time for a signal to travel from the call phone to each of several call turns in the vicinity of the user phone.

International mobile station equipment identity is specified by a kind of serial number allocated by the equipment manufacturers who stores it. This allows to locate stolen or obsolete non – functional equipment.

Long distance mobile calls.

The existing terrestrial networks are used to provide a microwave link between two locations. However, it is not possible to communicate long distances because terrestrial coverage is not available all over the world. Therefore, satellite are used for communication. By this method, the mobile phone can switch to satellites and provide worldwide connectivity to public. It is thus possible to provide connectivity between two remote places on earth by using satellites operating in Geostationary Earth Orbit (GEO). Sometimes a large costly terminal is required to make a phone call which results in high call charges.

The following types of orbits for mobile and other satellite systems are used;

i. Low earth orbit called polar orbit satellite (LEO)
ii. Medium earth orbit
iii. Highly elliptical orbit
iv. Geostationary earth orbit

Because of closer distance of low earth orbit satellites, the round trip propagation time is usually less than 20 micro second.

Global star system

It is another LEO satellite system. The call routing is done through large number of interconnected earth stations or gate way. This method provides public with localized regional phone number for hand sets. But when there are no gate way stations to cover some remote areas, facility of phone cannot be provided.

Cell phones operate at frequencies in the microwave range. Because they have short wavelength relative to traditional radio waves, smaller antenna are needed and the amount of data transmitted can be higher due to larger bandwidth. Microwaves are used to send voice signals from a mobile cell and also to receive them.

Communication method

Each cell phone contains a microwave transmitter and receiver (transceiver), antenna and control circuits. Our voice (sound waves) is converted into electrical signals (voltage) by piezoelectric components. This is then digitized. The digital signals is transmitted as a radio frequency signal by antenna in the phone. The cell phones also receive the electromagnetic wave and convert them into radio frequency signal in encoded form through the cellular network. It is converted back into sound by the speakers in the phone. The same antenna is used for transmission and reception. For this purpose special switch called duplexer is used.

Cellular mobile network

Cellular network consists of several such stations spread across the country. The base stations cover an area (range) for an individual cell. Data is transmitted from the phone to the base station and then to another phone. If the receiving station does not lie in the local call, the data will be sent to the base station of the cell where the receiving phone is located. The base station will send data to the receiving phone. At the receiving end, the RF energy is converted back into sound waves as mentioned earlier.

The cell phones use one of the two types of technologies called GSM or CDMA. GSM means global system for mobile communications. Cell phones use a cell phone service carrier GSM network with the help of cell phone towers. It was developed by the European telecommunications standards. It offers wider international roaming capabilities than other technologies and may be called a 'world phone'. It uses digital technology and is a second generation (2G) cell phone system.It is a cellular network which means that cell phones connect it by searching for cells in the immediate vicinity. Five cell sizes in the GSM network are used which are called macro, micro, pico, fento and umbrella cells. For example when the base antenna is installed on a mast or a building on roof top level, it is regarded as macro cells.GSM network operates in a number of different carrier frequency ranges from 900Mz to 1800 MHz bands in 2G system.

In the more advanced GSM standard a TDMA technology has been incorporated in 2 G cell phone system. It is an abbreviation of time division multiple access. This technology works by dividing each digital cellular channel into three time slots for increasing the amount of data carried. Multiple users therefore, share the same frequency channel without causing interference because the signal is divided into multiple time slots. Thus each conversation is transmitted alternately over short length of time with this method.

Another technology used in mobile phones is called code division multiple access (CDMA). It uses a spread spectrum methodology which is a wireless communication technique in which the users original signal is transformed into another form. This is done by changing to larger band width than the original signal. This process is called spreading. This allows many persons on multiple cell phones to be multiplexed over the same channel to share a band width of frequencies. CDMA networks operate at frequencies of 900MHz. In this system, data and voice packets are separated using codes and then transmitted over a wide frequency range. Thus allows multiple calls to be routed into the same channel. It permits several radios to share the same frequencies. In this digital radio system streams of bits can be transmitted. Since large number of phones can be served by smaller number of cell sites, CDMA based technology has a significant economic advantage over TDMA based standards or older technology using frequency division multiplexing.

In some cell phones two sim cardsi.e., both GSM or one GSM and the other CDMA may be used. Most mobile sets contain the following facilities:

i. Phone book
ii. Inbox and outbox for messages received and sent
iii. Video players and recorder
iv. Image viewer
v. Camera
vi. Sound recorder and player
vii. File manager
viii. Calculator
ix. Alarm setting
x. Games
xi. Bluetooth
xii. FM radio
xiii. E-mail facility

Sometimes, if there are problems in the reception of the signals in a mobile phone particularly if they are for away from a tower, a device called signal booster can be used. It uses a Yagi antenna mounted outside and then connected to mobile booster amplifier. However, the problems of non- functioning of mobile phone in a few basements of multistoried malls and other buildings need to be resolved by the mobile operators.

Mobile station

This term is used for a mobile phone for both transmission and reception of microwaves through a single antenna.

Base station (Microwave tower)

It is called a cell and is used to provide direct communication with mobile phones. When several such calls are grouped together, a cluster is formed. Each cell consists of small hexagonal area with a base station located at the centre of the cell. To establish communication between a mobile phone and a base station, two frequencies are required

i. Uplink- one frequency is used for mobile phone to base station (tower).
ii. Downlink- another frequency is used for communication from the tower to the mobile phones.

    Since microwaves travel in straight line but there is a loss of signal with the increase of distance, the geographical area is covered by several such cells. A cluster is formed by grouping these cells. In cities, each cell may work upto about a kilometer. It increases to about 8 to 10 km in rural areas. The range may further increase in large open areas.

    The microwave frequencies used in the transmission of signals are in the range of 800 to 2400 MHz. If the mobile station transmits at 825 MHz, the base station frequency will be kept at 870 MHz giving channel separation of 45 MHz. Each noise channel has about 30 KHz band widths. A simple calculation can be made to calculate the number of channels available per call for a particular cluster size.

iii. Mobile switching centre (MSC)
iv. Base station controller (BSC)

A group of base station (towers) are connected to base station controller. It consists of the following:

i. The electronics parts include radio frequency amplifiers, radio transceivers, radio frequency modulators, control and communication links and power supplies . It is located at the base of the antenna tower.
ii. The antenna and feeder mounted on mobile towers are connected to base transceivers.

iii. The location, height and orientation need to be selected carefully to cover the desired area. Otherwise, there may be a gap in the signal reception as could be with a lower antenna or there can be interference with another call using the same frequencies from a higher antenna.

The operation of calling from one mobile phone to another can be understood by using full duplex transmission systems. The method enables to talk or send data in both directions simultaneously. This is done by

i. Using two separate frequencies for the uplink and the down line as mentioned easier.
ii. A single frequency is used to transmit signals in both the directions.

In the former method, the information from the mobile phone to tower is carried on one frequency. The information from the tower to the mobile phone is received in another frequency. In the second method, information from the mobile phone to the tower is transmitted at one frequency and information from the tower to the mobile phone is transmitted at another time on the same frequency.

Two general types of channels called the control channels and traffic or noise channels are used. The former is also called paging channels in the case of downlink and access channel in the case of uplink. These set up calls coming from a mobile phone and then switched to voice channels.

The 3 G wireless phone booster is very effective to offset gaps in the mobile network and fills this blind zone to ensure communication. The frequency of 2100 MHz is generally used.

Cell phone standards

First Generation 1G; In theworld of cell phones, 1G signifies first generation wireless analog technology standards that originated in the 1980s. This was introduced as analog standard for cell phones.

Second Generation 2G; This signifies second generation wireless digital technology. It replaced analog 1G and was launched in 1991.

Third Generation 3G; This is the third generation of mobile phones technology. This has enabled faster data transmission speeds, greater network capacity and more advanced network services.It may be 5 to 10 times faster than 3G with maximum speed of 10 mega bits/sec.Watts App is used for quick transmission of photos, data and video through smart phones which use 3G technology.

Fourth Generation 4G; It is an improvement over 3 G with high speed wireless service in complete replacement for current network infrastructure and provides a comprehensive and server IP solution where voice, data and streamed multi media at much higher data rates than previous generations.

Superfast fifth generation (5G) wireless technology that would transmit high quality digital signals practically without limitations. The test showed transmission of more than one gigabyte per second over distance of two kilometers. The facility will enable 3 D movies and games, real time streaming of ultra high definition content and remote medical services.

The hazards due to microwave radiation from the towers and mobile phones have led to new set of guide lines by Government of India. These are discussed in Chapter 16.

Audio and video conferencing

Companies are often holding meetings and discussions from different locations using video conferencing. This is done via satellite communication. We can talk and see the photos of our distant relatives and friends or watch TV/video.

Space communications

The International Telecommunications Union has assigned a portion of the X band for deep space telecommunications. American NASA deep space network facilities are located in California, Australia and Spain. The stations also use S band deep space radio communication allocations. These three stations provide communications from the earth to almost any point in the solar system.

Military communication satellites

The International Telecommunications Union has assigned the X band uplink frequency (7.9 to 8.4 GHz) for sending modulated signals. The corresponding down link for receiving signals has been assigned a frequency band from 7.25 to 7.75 GHz. Both these have a band width of 500 MHz. The local oscillator frequency of X band low noise converter is 6300 MHz.

Killer satellites are designed to destroy enemy warheads, satellites and other space objects. The United States strategic command tracks space debris from disintegrated satellites to distinguish them from incoming missiles.

Microwave applications in Indian satellites

i. With passive sensors

   The first experimental remote sensing satellites built by India were called BHASKAR I and II. They carried microwave radiometers operating at 19, 22, and 31 GHz.

   IRS-P4 called Ocean SAT 1 had ocean colour monitor and multi frequency scanning microwave radiometer. For calibration, three sites were selected. Out of these, one site could be used as the microwave temperature measured by it was steady at only one site as compared to two other sites.

ii. With active sensors

   The first Indian active sensor on board Ocean SAT – 2 was Ku band scatterometer. Another satellite RISAT-2, radar imaging satellite was built by Indian Space Research Organization. It carried a synthetic aperture radar.

Indian national satellite system (INSAT)

This series of Indian satellites was launched since 1983 which were called INSAT satellites. The satellites are monitored and controlled by master control facilities at Hassan and Bhopal. These multipurpose satellites carry out the following:

i. Weather : Cloud location through photography, tropical cyclone detection and tracking, water vapour estimation. Reception and transmission of meteorological and other data from oceanic and land unmanned stations.
ii. Television and radio broadcasting
iii. Telecommunications
iv. Search and rescue operation.

This is achieved through microwaves, visible, infra red and water vapour channels carried by satellites as given below:

i. Visible (0.55-0.75 μm), thermal infra red (10.5-12.5 μm), water vapour (5.7-7.1 μm) channels with very high resolution radio meters (VHRR) ground resolution of 2x2 Km^2, 8x8 Km^2 and 8x8 Km^2.
ii. Camera for imaging.
iii. Normal C band transponders (12 to 24).
iv. Extended C band transponders.
v. S band transponders .
vi. Ku band transponders (6 to 12)
vii. Data relay transponder with global coverage for reception, with 400MHz uplink and 4500 MHz downlink.
   Other Indian satellites
   Kalpana -1: Exclusive meteorological satellite with VHRR and data relay transponder for meteorological services.

INSAT satellites are a joint venture of the Department of Space, Department of Telecommunications, and India Meteorological

Department. In addition to serving the needs of communication, meteorological imaging and television, the satellites incorporate transponders for receiving distress signals for search and rescue missions in the south Asian and Indian ocean region. All these geostationary satellites are generally located near longitude 74^0 E or 93.5^0E or 48^0 E.

Most of these satellites have been launched by a vehicle called PSLV.

GSAT

Somewhat similar to INSAT 3 B and 3C for high power communications purpose, it has several C band and extended C band transponders. G SAT 8 also carries two channel global positioning system (GPS), aided geo augmented navigation (GAGAN) in L1 and L5 microwave bands. GSAT 10 has 20 transponders in Ku band, C band and extended C band. It has navigation pay load GAGAN for improved accuracy of GPS signals for civil aviation requirements. G SAT 12 augments the capacity of INSAT for teleeducation, telemedicine and for village resource centre.

Fig. 8.2 GSAT-7 or Rukmini at its launch in French Guiana

EDU SAT

It was launched to meet educational requirements. It carries Ku band transponders and extended C-band transponders.

Indian remote sensing satellites (IRS)

IRS satellites as earth observation satellites were launched since 1979. Data from the IRS satellites is received and disseminated by several countries in the world. The satellites are placed in polar sun-synchronous orbit with varying spatial and temporal resolution. The data is useful for resource survey and management in the following broad areas:

i. Agriculture like crop production estimates, drought monitoring.
ii. Water resources like flood damage assessment, hydro geomorphologic mapping for underground water resources, irrigation command area, monitoring snow melt run off estimates.
iii. Forestry and ecology – survey and environmental impact analysis.
iv. Geology-mineral prospecting, mapping of fault zones causing earthquakes and tsunami (sea waves reaching enormous heights near coasts and gushing water with very great force generated by earthquake) oil exploration, landslides.
v. Coastal studies like oil pollution, sea waves height, winds over coast, sea surface temperature, ocean geoids.

The initial IRS satellites were named as IRS 1 A to IP. But the later versions were classified by keeping in view the area of application like Ocean SAT, Carto SAT and Resources SAT. It may be clarified that Resources SAT has advanced wide field sensor and atmospheric correction senser for quantitative retrieval of geophysical parameters.

Carto SAT 3 has resolution of 30 cm and 6 km swath for infrastructure mapping and disaster monitoring and damage assessment . Ocean SAT 3 is meant for ocean biology, sea state analysis, and potential fishing zones. In addition to thermal IR, ocean colour monitor, it will carry scatterometer and passive microwave radio meter.

Satellite data reception and dissemination

The National Remote Sensing Centre at Hyderabad not only acquires and processes data from Indian satellites but also receives data from foreign satellites like Terra, NOAA and others.

CHAPTER 9

Microwaves in geodesy and navigation

The development of Global Positioning System (GPS) using microwaves has revolutionized navigation and geodetic observations. The latitude and longitude of a place were computed earlier from base line observations using triangulation and trilateral methods. This aspect and shape of earth is included in 'geodesy'. In the earlier days, the common instrument for navigation over sea, land and air used to be a magnetic compass showing the direction with respect to North. Radio aids were also utilized for navigation system such as LORAN which were used during World War II. Satellite navigation in recent years has become the backbone of modern civilization. The microwave based System called Global Positioning system (GPS) used 24 satellites by USA initially for military purpose (Fig 9.1). It was allowed access for civilian usage worldwide. The Russian Global Navigation Satellites system (GLONASS) was also developed at the same time but had limited coverage of the globe. Satellite navigation has now become a global program. The use of GPS is common in many of scientific and navigational activities to the extent that it can drive a car to destination without knowing the roads. GPS gives the latitude longitude of the place, its height and the time of observations. The accuracy of these measurements is generally less than 1 cm. GPS consists of the following:

i. Satellite in orbit with precision clock transmits signals continuously in time when the message is originated. The oscillators in satellite generate frequencies of 19 cm and 25 cm (L band).

ii. GPS antenna and receivers which receive signals from the GPS satellites. The heart of each receiver is its oscillator which drives the clock. The receiver determines the transit time of each message. It then computes the distance to each satellite using the speed of microwaves ($3x10^8$ m/sec). Different codes are available for civilian and military purpose.

These distances and satellite positions are used to compute the location of the receiver by using mathematical equations. It is necessary that each receiver makes measurement from several satellites. The positions of the receiver can be shown on a map or can be printed as latitude, longitude and altitude. In practice, four or more satellites are needed for accurate location. The additional satellites ensure considerable redundancy to improve accuracy over the minimum of satellites needed for a location (position). Generally, 8 to 10 satellites are visible from a given point on the earth's surface.

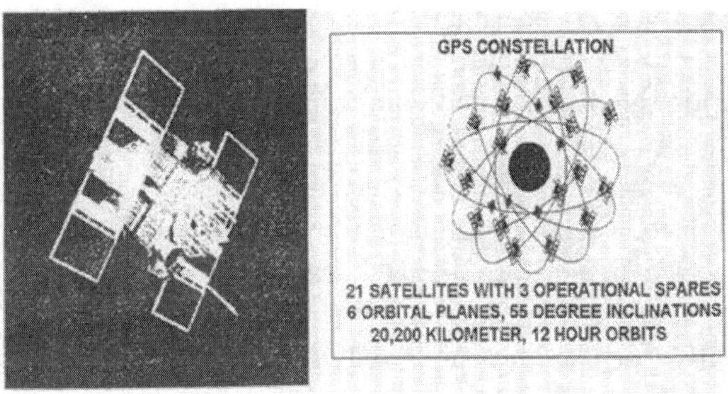

Fig. 9.1 Global Positioning System

GPS satellites are kept in orbit at a height of 36,000 km and repeat the same ground track each day. Each receiver gets signals from many satellites which are affected by the following errors:

i. Error in satellite clock
ii. Error in receiving clock
iii. Tropospheric effects on time delay
iv. Ionosphere effects on time delay

Errors due to receiver clocks are removed by observation from two satellites. Similarly, errors of a satellite clock are removed by two receivers.Since microwaves travel through troposphere and ionosphere, both of them introduce some delay in the time of propagation. Tropospheric delay errors are corrected by weather observations using atmosphere pressure, temperature and humidity from which refractive index of air is calculated using the MET Pack provided with each GPS. Radio signals refracted by the ionosphere depends upon the total electrons content and the signal frequency. Signals monitored at two frequencies using dual frequency GPS receivers therefore, provide a method to estimate the time delay by the ionosphere. The path delay caused by the errors due to solar corona and inter planetary plasma are removed by using the multi frequency link between the probe and the ground station. All such systems apply CDMA code for measuring the distance and data transfer. The signals are spread spectrum modulated on the L1, L2 and L5 bands. CDMA offers many advantages in respect of accurate measuring the time of arrival of the signals rejecting multipath and simple decoding of the received signal. Also all satellites can broadcast in the same part of the spectrum.The troposphere time delay is also utilized to estimate total water vapour and rainfall in the atmosphere and will be discussed in Chapter 10.

To monitor the area for earth deformation a number of points that are monumented are used. Of these, some stations are used to define a stable reference frame. The remaining stations are the monitoring points situated in the deformation area. In this way, the determination of control stations is done relative to the reference stations. This method enables us to measure the movement of the deformed area. The applications of GPS in the earth's plate motions, land slides and glacier studies are discussed in later chapters.

Applications

GPS is now a days widely used for the following applications:

i. Clock corrections: The accuracy of the GPS time signals is almost same as atomic clocks.
ii. Navigation gives precise location in ships, cars, trucks, aircrafts.
iii. Surveying helps in preparing absolute locations which enable to prepare maps along with their boundaries.
iv. Mobile phones
v. Earthquake studies- plate motions, fault parameters
vi. Volcanic eruption warning
vii. Weather forecasting – estimation of water vapour and rainfall
viii. Space research

The Survey of India is bringing out road maps using GPS and other survey methods.

GPS observations along with tide gauge records near coastal stations help in estimation of absolute sea level rise which is essential to estimate the influence of global warning. GPS application on cellular or mobile phones has the advantage that it can be used as a means of communication along with camera with high resolution to take photos. These give coordinates which can be sent directly to any place or office for on-sight monitoring. GPS enabled phones help to find nearby restaurants, fastest route to home, keep track of children or elderly persons with a tracking service.

Global navigation satellite system (GNSS)

This system includes GPS (USA), GALILEO (Europe) and other regional satellite navigation systems. The capability of GNSS to provide very accurate position and time information in all weather conditions has made it a valuable means for civilian and scientific applications. GNSS comprises of

i. Satellites components (24 USA satellites, 21 Russian satellites and others)
ii. Ground or terrestrial control equipment
iii. User equipment
iv. Propagation media

GNSS has following applications in earth science:

i. Earthquake studies,
ii. Atmospheric studies
iii. Land slide monitoring
iv. Volcanic eruptions
v. Glaciology
vi. Geoid determination giving geoid height at a point.

GNSS vulnerabilities are due to positioning errors caused by reduced availability of satellites GNSS signals, multi paths and receiver noise errors and propagation media (ionospheric and tropospheric induced). Also geomagnetic and ionospheric storms pose threats which are caused by solar disturbances and can last from several hours to several days. The distribution of charged particles in the ionosphere can also be affected by volcanic eruptions. If GNSS performance deterioration lasts for a longer time, corrective actions need to be taken like closing sub systems temporarily and utilization of alternative systems.

GAGAN

This is a joint project of Airport Authority of India and Indian Space Research Organisation. It is a satellite based augmentation system to the GPS (GAGAN). It is used to augment the GPS signals with the necessary correction for meeting the accuracy specified for air craft navigation consistent with International Civil Aviation organization standards(Fig 9.2).

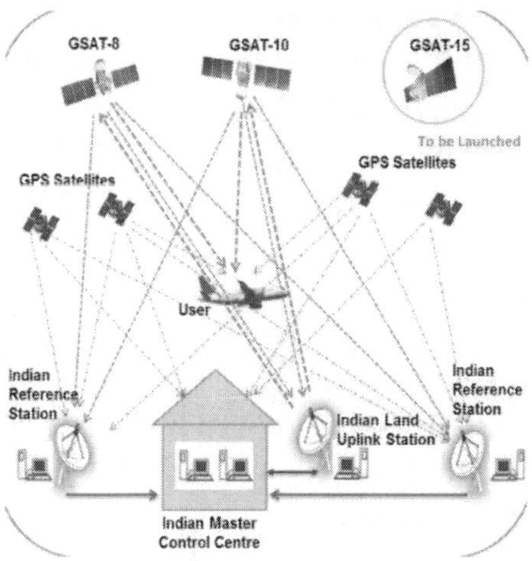

Fig. 9.2 – GAGAN – Venture of ISRO

With the help of 7 satellites, it provides all weather navigation in the Indian land mass and beyond its geopolitical border. In GAGAN, corrections to GPS due to ionosphere error, clock error and ephemeris error are provided by the ground segment by a reference station using S and L microwave band and besides CDMA laser ranging of the satellite. The L band spectrum for radio navigation satellite spectrum for radio navigation satellite service is very crowded. However, GAGAN signals for non aviation users are available for survey, road and rail applications, vehicle tracking, traffic systems, land management, maritime uses, geodynamics and research.

International GPS Service (IGS)

The service is based on about 150 globally distributed permanent GPS tracking sites, global data centers and regional data centers. In India, such station is located at Bangalore.

IGS routinely provides;

i. High quality orbits for all GPS satellites.
ii. Earth rotation parameters.
iii. Contributions to the determination of the tracking site coordinates in the international terrestrial reference frame.
iv. Phase and pseudo range observations for each IGS tracking site.
v. Ionosphere information
vi. Tropospheric information

Increasing volume of traffic leads to traffic congestion, environment pollution and traffic accidents. Many countries like USA, European Union and Japan are using GNSS to enhance safety and efficiency of transportation. The system is not only used to determine vehicle's position but to gather information on goods transfer. In India the use of GPS in all types of vehicles will help in tracking major crimes such as kidnapping, rape, murder, thefts of goods or their disappearance during transportation.

In a localized area within a country, local navigation satellites systems are useful. For the transportation of goods, food and industrial products between countries, a secure and safe tracking system is required. A global system like GNSS may be used for tracking on different routes. This is particularly useful to track hijacking of ships over oceanic regions.

New GPS technology in cockpits of aircrafts is a part of next Gen Air Traffic control system. So far, ground based radar control system is being used for Air Traffic Control. The new GPS based microwave system is expected to improve safety, reduce air traffic congestion, increase traffic capacity and shorter flight times.

If the data are acquired by different set of transmitters, they have different time references like GPS time or Galileo time. It is, therefore, necessary to take into consideration difference among the different time references.

Care has to be taken that GPS systems are not faulty otherwise wrong directions will be shown which may cause accidents. Also GPS spoofing of fighter planes can be disastrous during wars.

GPS Jammers

GPS jammers send very powerful signals to obstruct the functioning of GPS receivers and satellites. Some companies intentionally send them to cause harmful interference. Sometimes the effect may be temporary. Different spoofing techniques can modify the actual broadcast signal. GPS can also be misused for illegal surveying and mapping.

Geodesy

Geodesy is a discipline which deals with the positioning measurement of earth, its shape and gravitational field. It also includes study of geodynamic phenomenon such as crustal motion, tides and polar motion. The positioning requires a frame of reference.

Point positioning is the determination of the coordinates of point on land, sea or in space. They can be measured by triangulation method or GPS satellites. For height determination, GPS requires precise information of the figure of geoid. GPS gives heights above the reference geoid.

The other space techniques for making precise position measurements between sites on earth surface separated by thousands of kilometer use very long base time interferometry

(VLBI) and satellite laser ranging (SLR). VLBI depends upon the precise timing of radio noise from extra galactic sources which are observed by several radio telescopes. In this method, two or more telescopes simultaneously record noise. The time delay between different locations are used to determine site locations. Experiments need to be repeated to estimate the motion of the site.

Satellite laser ranging is based on the round trip time of laser beam reflected from a satellite. Range estimates for many sites determine both the satellite trajectory and the coordinates of the tracking station. Successive observations are taken to measure the movement of the station. When multiple satellites are used, the location of the station can be estimated to within a few meters by trilateration.

Car accident prevention

A car radar system can detect cyclists and automatically apply the brakes to avoid fatal crashes. It can also sound an alarm to the driver. The radar system is fitted in the cars with a camera behind the rear view mirror to detect possible collisions. GPS monitors the position of the car.

It would thus be seen that maximum benefit of GPS can be gained through international cooperation. In the words of Dr. B.W. Parkinson, the chief architect of Global Positioning System, "The advantage of GPS based weapon delivery is that if the target is accurately located, you hit the target; you do not hit the nearby mosque, or church, or temple or hospital or whatever. He said that GPS is a humanitarian weapon system."

CHAPTER 10

Microwaves in weather forecasting

Weather forecast helps in planning our activities and agriculture which has direct impact on national economy. With the advent of weather radar, satellites mounted with microwave instruments like scatterometers, GPS, microwave imagers, there is a phenomenal increase in meteorological data which has a direct impact on the quantitative aspects of weather forecasting.

Weather forecasts are issued for different time frames ranging from nowcasting, (a few hours), short range (one day, outlook for next 2 days), medium range (7 to 10 days) and long range (month to season). The accuracy in the first two types of forecasts has tremendously increased with the weather satellites and microwave radars. Modeling studies and satellites can give a lead time of a few days as happened during the severe cyclonic storms, Phailin(2013), Hudhud (2014), Vardah(2016), Ockhi(2017), Phethai(2018), Fani(2019), Amphan(2020), Tauktae(2021). The long warning time of more than 2 days helped in timely evacuation of people from the danger zone to cyclone shelters resulting in very few casualties.

Weather forecasting is primarily based on meteorological instruments which record surface weather observations every three hours. These instrument are kept in a ventilated chamber called Stevenson screen at a height of about 1.7 meters above ground. The maximum and minimum thermometers and dry bulb and wet bulb thermometers (to measure humidity) are kept inside the screen. Other instruments like wind vane and anemometer are used to measure wind speed and direction. They are mounted on the top of building to measure undisturbed flow of air. The amount of rainfall is measured by rain gauge and is kept on the ground. Atmospheric pressure is measured by a barometer kept inside a room. Cloud observations and visibility are also reported. These observations are sent to the central recording stations by FAX, internet and telephone. Nowadays, automatic weather stations are increasingly being used. The observations of soil radiation, leaf wetness, soil moisture and temperature are also taken at some places which are utilized for agricultural purpose. All the data is communicated by satellite link, USB port, LAN, WAN, GPRS, GSM, CDMA, dial up modem and internet.

Over the oceans, the data buoys and ships send meteorological data via observational satellite. GPS met package allows to compute integrated water vapour profile with height at the place. This is validated by radiosonde instruments flown in balloon containing temperature, pressure and humidity sensors. The data is continuously transmitted to the observatory using microwaves to receive and record pressure, temperature and humidity at different heights in the atmosphere.Wind speed and direction is also recorded by a special type of radar. High altitude cloud physics laboratory has been set up at Mahabaleshwar at an altitude of 1380 m in Maharashtra near Pune. This observatory has microwave radar and microwave radiometer besides other instruments to study cloud condensation, aerosol and precipitation properties at high altitude.

In India, global meteorological observations are being received through the global telecommunication system (GTS) of the World Meteorological Organization (WMO). The observations received via GTS along with other satellite observations downloaded through internet are then processed at National Centre for Medium Range Forecasting and India Meteorological

Department. This data is used for preparing weather charts at synoptic hours for issuing weather forecasts. They are also utilized in numerical weather prediction models, which give forecast for different periods. Weather forecasts are required by Government, public, agriculturalists and numerous other applications.The economy of our country also depends upon the rainfall during the southwest season called the rainy season.

The spatial and temporal variation of soil moisture is needed for crop yield models, crop health and dry land farming. Microwave sensing is used for estimation of soil moisture because of its larger penetration capacity as compared to optical waves and also because they are sensitive to the presence of moisture in the soil. As explained earlier, microwave propagation depends upon dielectric constant of the material it strikes. Due to large difference of dielectric constant between dry soil (2 to 3) as compared to pure water (80 at room temperature and at about 1GHz), the dielectric constant will vary between these two limits. As dielectric constant also depends upon frequency, soil moisture upto deeper layers can be estimated by using high frequency microwaves. However, in the areas with very large amount of water, the absorption will be rapid and microwaves can estimate soil moisture to very shallow depth.

Weather changes from day to day and time to time during different seasons called diurnal and seasonal changes. However, large scale changes in weather like rainfall, hail, strong winds occur with moving weather systems. The western disturbances are low pressure systems which move from Mediterranean Sea through Iran/Iraq, Afghanistan to northwest India. These may cause snowfall over hill stations mostly during winter months and rainfall in northern India. A few weather systems which move from east to west commonly affect peninsular India. Heavy to very heavy rainfall occurs in association with monsoon depressions which form over Bay of Bengal and sometimes over

Arabian Sea. During the months of April, May, and October to November, very violent low pressure weather systems called tropical cyclones form over the Bay of Bengal and Arabian sea when the sea surface temperature reaches 28 ^{0}C or more. Although monsoon depressions also cause very heavy and continuous rain in its passage over the inland stations, the tropical cyclones are most damaging due to;

i) Violent winds of 150 to 250 km per hour.
ii) Heavy to very heavy rainfall
iii) Storm surges, 3 to 5 meters high rushing in land and sweeping off everything in the way causing large loss of life and destruction to property. Heavy rain causes flooding and further damage.

The cyclonic storm is a cylindrical column of winds which has a diameter of about 600 to 1000 km rising to about 10 km in the atmosphere. In its mature stage, a calm central area (wind speed variable or very light or calm, very light rain or almost clear skies) of diameter between 10 to 50 km develops. This is called eye of the tropical storm. Highest winds and heaviest rain occur at the periphery of the eye which decrease outwards.The wind speed is the main criteria to categorize cyclonic storms of different intensity. A weather system with wind speed of about 50 km per hour is defined as a tropical cyclone. The highest category super cyclone has a wind speed of 220 km/h or more. The cyclones are called hurricanes in Atlantic and typhoons in Pacific.

Radar observations of tropical cyclone.

The radar at a coastal station detects cyclonic storms from the spirally shaped cloud pictures. The radar also shows the diameter of the eye, its shape, height of cloud tops, number and length of spiral brands. Doppler radar shows that in a hurricane, maximum wind and the maximum reflectivity both sloped radically outward with height. Maximum rain also coincides with the radius of maximum winds.Although tornadoes have not been reported from the tropical cyclones in the Indian region, they have been documented in the Atlantic basin. They are detected by Doppler radar which shows height reflectivity maximum.

Radar also helps in tracking the motion of tropical cyclones. The centers are plotted at regular intervals of time and give the direction in which the cyclonic system is moving. These observations combined with satellite are used to estimate the speed of the tropical cyclone (Fig.10.1).

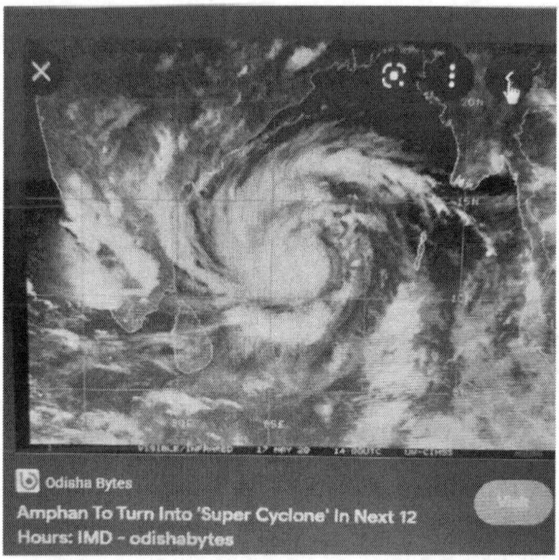

Fig 10.1 Satellite observations of severe cyclone Amphan(2020)

Doppler weather radar of 10 cm wavelength is also used for;

i. Severe weather monitoring
ii. Hail detection
iii. Flood forecasting
iv. Airport wind shear detection
v. Weather modification.

Satellites for weather forecasting

After seeing the utility of TIROS and other satellites by USA and other countries, INSAT series of satellites launched by India have been regularly keeping a track about formation of cyclones and other weather systems overseas as well as inland.

The tropical rainfall measuring mission (TRMM) is a joint space mission between NASA and the Japan Aero Space Exploration Agency. This mission is accomplished through a satellite to monitor tropical rainfall. The following instruments are located on the satellite.

i. **Precipitation radar**
   Through three dimensional images of storm structures, the radar measurement gives information on the distribution of rain, its intensity and the depth of the storm.

ii. **Microwave imager**
   It is a passive microwave sensor to provide quantitative rainfall information over a wide swath under the satellite. This gives quantity of water vapour, cloud water and the intensity of rainfall in the atmosphere.

iii. **Visible and infrared scanners**
   These measure radiation in five spectral regions from 6.3 to 12 micrometers. They give a first hand information about areas of rainfall.

iv. **Cloud and earth's radiant energy sensor.**
   These measure the energy at the top and within the atmosphere and at the earth's surface. High resolution imaging instruments detect cloud amount, altitude, thickness and size of cloud particles.

v. **Lightning imaging sensor**
   This instrument detects and locates lightning over the tropical region and also the flashing rate.

Scatterometers are satellite borne active microwave sensors to determine the wind direction and wind velocity over sea(Fig.10.2).They send out a signal and measure the amount of signals returned after interacting with the target. Microwaves are scattered by short water waves. The portion of energy returned as a back scatter to the satellite depends upon wind direction and wind speed. The wind speed is determined from the strength of the back scattered signal. The wind direction is found by determining the angle consistent with the back scatter observed from multiple angles roughly for 5 minutes. Mathematical functions are used for this purpose. These observations are then fitted as gridded products from other observations. They are uscd in weather models to fill gaps in data over oceanic region.

These instruments are useful for

i. Weather forecasting
ii. Marine safety
iii. Commercial fishing
iv. EL Nino prediction which has world wide effect on weather.
v. Long term climate studies

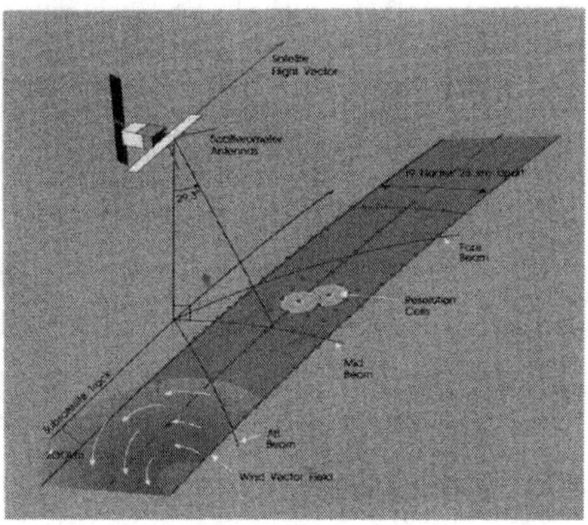

Fig. 10.2 Wind Scatterometer aboard ERS-1

Telecommunication networks have a promising potential to provide rainfall estimates that are complementary to those from raingauges and radar networks. The attenuation of microwave signal at the frequencies used by them is nearly proportional to the path averaged rain rate.

Climate Change

Global warming is the increase in the average global temperatures over time. It is caused by a combination of natural and human factors which increase the greenhouse effect of gases, mainly carbon dioxide, due to the combustion of fossil fuel and deforestation. This results in the rise of atmospheric temperature. Due to this, glaciers are melting and sea level may rise. Thus, the life of people living near the coasts and over the islands is threatened and they may need evacuation. However, reliable data on the sea level variations is needed. For monitoring sea level changes tide gauge and radar altimeters aboard satellite are used.

Measurement errors on mean sea level from the missions like SEASAT (1978), GEOSAT (1985 to 1989) and ERS-1 (after 1991) were large projects undertaken to detect small changes in global mean sea level.The main errors were noted in determining the satellites altitude, the ionosphere delay correction, the wet troposphere delay correction and the long term calibration needs. However, the US-France TOPEX/POSEIDON satellite altimeter mission, launched in 1992, showed considerably improved results in sea level changes due to reduction of many earlier errors of measurements. There are two radar altimeters on board the new satellite. One is a dual frequency altimeter which operates 90% of time and a single frequency altimeter which operates during the remaining time. The satellite altimeter mission measured global men sea level every 10 days for 2 years with a precision of 4mm approaching the requirements for climate changes research. It was, however, found that data for longer time is needed for climate changes studies.

CHAPTER 11

Microwaves in earthquakes and tsunami warning

Earthquakes have been causing huge destruction and loss of lives throughout the world. India also lies in seismically active region due to the north-northeasterly movement of Indian plate and its collision with the Eurasian plates causing great faults in Himalayas. Great earthquakes have occurred in Himalaya and northeast India in 1897, 1905, 1934, 1950 and 2015 which caused damage and loss of life over very large areas. Recurrence of major earthquakes (magnitude 7.6-7.8) in1819 and 2001 in Kutch region within the Indian plate have been attributed to multiple sources in the rift zone. The tsunamigenic earthquake of December 2004 (magnitude 9.2) in the Indian ocean with its epicenter near Indonesian islands caused the death of about 2 lakh people in coastal regions of India and other coastal countries in southeast Asia.

Microwave communication for earthquake studies has been utilized towards the following objectives.

i. Determination of epicentral parameters of earthquakes in near real time by use of V- SAT for data transmission and reception.
ii. Tsunami warning system
iii. Early warning system for earthquakes
iv. Disaster management

The epicentral parameters of earthquakes (place or epicentre, focal depth and magnitude) are required by the Government, press and others in the shortest time which has now become possible with modern microwave communication systems (Fig 11.1).Towards achieving other objectives like earthquake hazard assessment, microwave instruments like ground penetrating radar, GPS (GNSS) and synthetic aperture radar are also deployed in addition to seismographs and other instruments. Ground penetrating radars are used to identify the fault along with other conventional geophysical and geological methods. The fault parameters which are determined by recording of seismic waves by the seismographs are now supplemented by synthetic aperture radars.

Although no reliable technology for prediction of earthquakes is available, very long time base interferometers, synthetic aperture radar and GPS are used to study crustal deformation which may precede earthquakes in some regions. The rapid microwave data transmission and reception methods have enabled us to design tsunami warning and early earthquake warning systems which can save millions of lives.

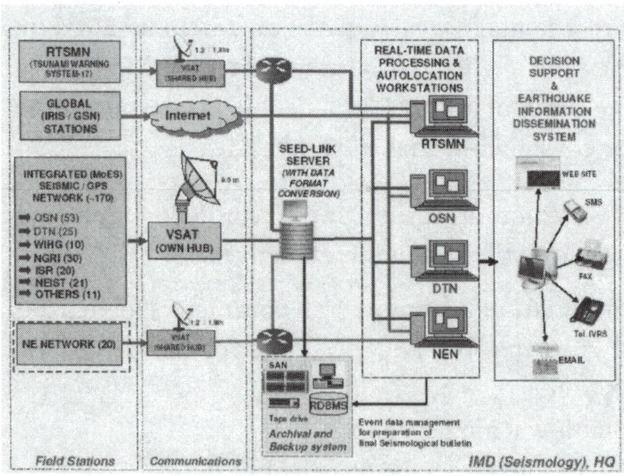

Fig. 11.1: Seismic telemetry system and data management at the National Centre for Seismology Ministry of Earth Sciences

Earthquakes are caused by weak zones called faults. The concentration or belts of earthquakes in specific zones was explained by the plate tectonics theory (Fig.11.2). According to this hypothesis, great earthquakes occur at the boundary between two plates. The whole earth is divided into six major plates. The earthquakes in Himalaya are attributed to collision of India and Eurasian plates. The north northeasterly movement of the Indian plate moving at the rate of about 5 cm per year accumulates enormous stresses near its boundary. When these stresses exceed the bearing capacity of rocks, they slip generating seismic waves to cause earthquakes. The movement of the plates can be monitored by a network of global positioning system (GPS) which communicate with the network of satellites as explained earlier (Chapter 8.1).

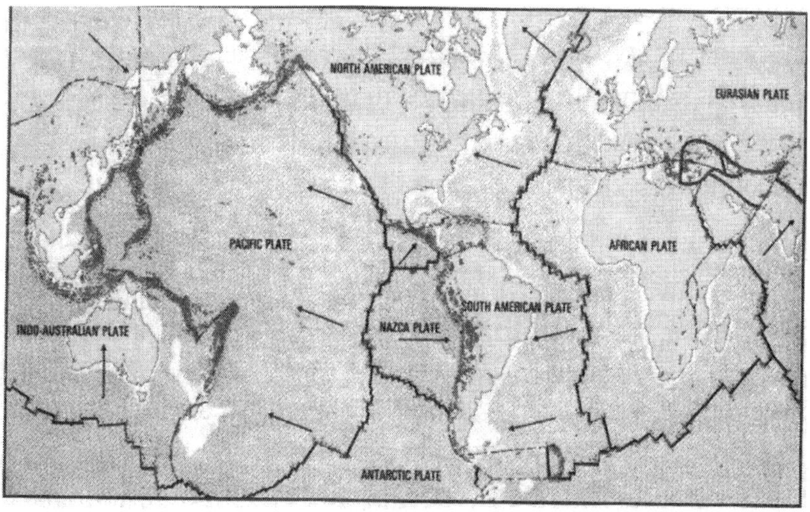

Fig.11.2 Major plates and epicentral belts(USGS)

Destruction due to earthquakes occurs because body and surface seismic waves from the source of earthquakes propagate in all directions resulting in acceleration or a force on houses, bridges, and other structures. The damage due to earthquakes is assessed by geologists, seismologists and earthquake engineers. The scale of damage is described in terms of seismic intensity called modified Mercalli scale (now MSK) expressed by I to XII Roman numbers. The damage in this scale starts with intensity of IV or V. The damage is total for intensity XII and nothing remains in the area effected by the earthquake of such a massive scale. The seismic intensity decreases with the epicentral distance being larger on soft or muddy or wet ground as compared to hard rock and also depends on the focal depth of the earthquake. But magnitude of an earthquake is calculated from the records of earthquake called seismogram. It is a fixed quantity which gives the energy of an earthquake. Another phenomenon called liquefaction occurs near rivers, lakes and very shallow water table during strong earthquakes. Due to the intense vibration during great earthquakes, solid soil gets mixed with water close to river banks or filled up ponds. The solid earth is turned into liquid and therefore, buildings will either be tilted or destroyed. Sometimes, sand blows may also form in these regions. GPR system has been used by the University of Baroda to image two large such blow craters which formed during 2001 Bhuj earthquake, Kutch, Gujarat. GPR data showed a sharp contrast in the sediment character between the inner and outer side of the crater. It was concluded that the large sand blows during 2001 Bhuj earthquake were produced due to liquefaction of sediments in the sub surface at depth larger than 6.5 m.

Earthquake monitoring

The Centre for Seismology, Ministry of Earth Sciences is the nodal agency of Government of India for monitoring seismic activity in and around the country. Currently a seismological network, consisting of about 100 observatories, spread over the entire country is maintained. This includes:

a. 16 station V-SAT based digital seismic telemetry system around National Capital Territory of Delhi
b. 20-station VSAT based real time seismic monitoring network in north east region of the country and
c. 17 station real time seismic monitoring network to monitor and report large magnitude under sea earthquakes capable of generating tsunamis on the Indian coastal regions. The operational task is to estimate the source parameters of land and under sea earthquakes capable of generating tsunamis, immediately on their occurrence and disseminate the information to all the user agencies such as public information channels, press, media and concerned state and central Government agencies responsible for carrying out relief and rehabilitation measures. The earthquake information is transmitted using various modes of communication, such as SMS, Fax, Email, IVRS and is also posted on the website.

Mapping topography and active faults

Topographic maps (1, 50, 000) are prepared by Survey of India. High resolution CORONA KH-B satellite data provides a ground resolution of about 2 m, while Shuttle radar topographic mission gives resolution of about 90m. Higher resolution satellites are now deployed for this purpose.

Fault mapping

GPR surveys have been carried out in Gujarat, Delhi, Chandigarh and a few other places in India to understand the nature of faults which cause earthquakes.GPR data help us to identify the nature and geometry of the tectonic structures below the ground surface as they give a high resolution continuous image. Some results using GPR survey (200 MHz) along Kachchh mainland fault to map its near surface trace and the geometry of the fault are shown in Fig 11.3. Strong reflections from the fault plane are clearly seen in GPR profile.

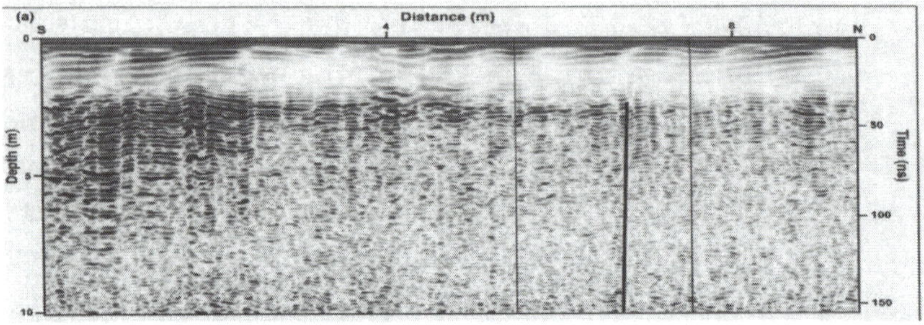

Fig. 11.3 GPR profile with 200 MHz antenna showing strong reflections from the fault plane. (Ministry of Earth Science report by University of Baroda)

The image brightness of the synthetic aperture radar depends upon radar back scatter and pixel level. It is a function of the dielectric constant, slope, surface roughness and surface discontinuity. Their application to several destructive earthquakes has enabled to map crustal deformation changes and dimension of faults. The procedures for determining earthquake source parameters from radar interferometer data are well established. These are compared with other models of the earthquake fault parameters based on seismological data.The phase in the seismic interferogram is more sensitive to the surface displacements than to the topography. Thus the former displacements contribute substantially to the interferometer fringes. The topographic signal is removed by comparison with pre seismic interferogram. The resulting interferogram gives surface displacement relative to earthquakes and those of aftershocks.Combination of pairs of synthetic aperture radar images acquired by ERS-1 satellites before and after the California earthquake of May 17, 1993 mapped the coseismic deformation. Their composite interferogram resulted about 10cm of change in range indicating a coseismic displacement due to deepening of the Eureka valley graben. These estimates indicated normal faulting on a rupture plane. This was in agreement with the distribution of aftershocks and mapped faults.

GPS studies are being carried out worldwide to find the movements across plates and local faults which cause earthquakes. In Japan more than 1, 000 GPS stations are in operation. Crustal movements at GPS stations were studied before and after the great east Japan earthquake (M 9.0) on March 11, 2011 which killed about 20, 000 people mainly due to the giant tsunami of 10 to 30 meters in height. More than 2 cm abnormal signals were noted in height at Ojika on 25 February and 8 March 2011 before the earthquake. GPS has also shown pre seismic signals of Muzaffarabad, 2005 earthquake in daily distance change ratio networks as well as daily triangle area change ratio 2 to 5 days in advance. The network of GPS stations in India is being expanded. The coordinated project under the Ministry of Earth Sciences for GPS studies has corroborated the north eastern movement of the Indian plate as a rigid body. The local variations of horizontal velocity of deformation near Indian plate boundaries are not yet conclusive for assessing impending earthquakes. GPS observations have also given valuable information about pre seismic and post seismic deformation and the extent of locked zone which may give an assessment of the largest magnitude of earthquake.

Of late, satellite images of earthquake stricken region have been of great help to victims and identify damaged roads and infrastructure. High resolution satellite images enable government to cope with natural disasters for organizing rescue and recovery options.

Tsunami warning system

Tsunami is a system of ocean waves formed as a result of large-scale disturbance of the ocean floor. These waves of a few tens of centimeter height travel at the speeds of jet plane in the deep ocean. They grow to large amplitude waves as they arrive at the coast due to conversion of kinetic energy of waves in the ocean into potential energy. This causes large scale inundations and destruction. The cause of tsunami are under sea earthquake, under-sea landslide, under sea volcanic eruption or explosion impact of meteors (Fig.11.3).

Earthquakes cannot be predicted but tsunami can be detected well in advance from a few minutes to hours depending on the distance of the epicenter of earthquake. This is because seismic waves travel much faster (6 to 8 km/sec) than the speed of tsunamis (200m/sec). Thus epicenters of earthquakes if determined within a few minutes and have magnitude of more than 6.5 can have potential to generate tsunami provided their focal depth is shallow and has occurred in subduction zones of the two plates. Tsunami mitigation can be effective by an appropriate warning system to alert coastal communities that are vulnerable for tsunami (Fig 11.4).

Basic components

Warning Centers use real or near real time seismological and water level data to rapidly locate the possible source of tsunamigenic earthquakes with a high degree of confidence.

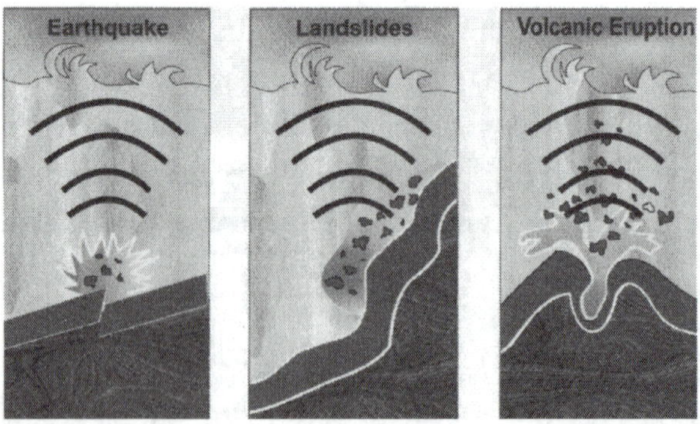

Fig. 11.3 Causes for Tsunami generaton

i) Earthquake data
Seismic waves which propagate rapidly through the earth are used by warning centers to detect the size and location of an earthquake. This data is sent to the centers in real time to determine the epicenter of an earthquake. Data from dense array of seismic stations surrounding nearby earthquakes is often used to get a quick and accurate location.

ii) Sea level data
Water level gauges are used to confirm the existence or non-existence of tsunami waves due to the earthquake. The water level gauges which are used for tsunami warning are specially designed to determine the small change in water level.

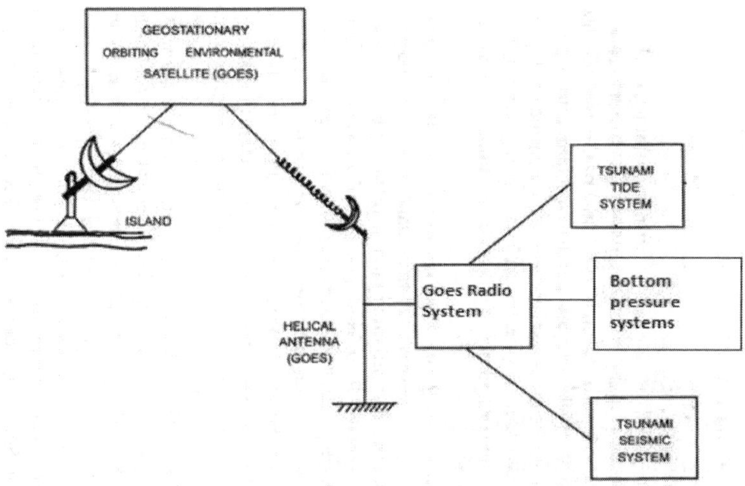

Fig. 11.4 Tsunami warning system

Deep ocean assessment and reporting of tsunamis (DART)

New technologies have been developed to improve the detection of tsunami, an automated earthquake detection and tsunami evaluation system, through the DART system. This is a data buoy which can detect even a small difference in the ocean height. This device measures the pressure difference caused due to the movement of tsunami. Wind waves which have a much shorter period are filtered out. If tsunami is detected, data is updated from an onboard computer chip via sound signal, to a hydrophone which is attached to the surface buoy. The buoy sends the data to the Geostationary Operational Environmental Satellite (GOES) orbiting the earth. The information is received from satellites to the ground based computers for immediate transmission by NOAA's tsunami warning system after processing (Fig 11.4). It has been suggested that GPS methods may be an improvement over seismic methods to compute the magnitude of tsunamigenic earthquake.

International and Local warning system

An International cooperative effort was taken to save lives and protect property in the areas affected by Tsunami.

Local tsunami warning systems have been developed in Japan, France, Russia, Chile, India and USA. The Tsunami Warning System (TWS) in the Pacific comprises 26 participating international member states. The objective of the PTWS is to detect, locate and determine the magnitude of potentially tsunamigenic earthquakes occurring in the Pacific Basin or its immediate margins. If the location and magnitude of an earthquake meet the criteria for generation of a tsunami, a tsunami warning is issued to warn tsunami hazard. The warning includes arrival times of tsunami within the area defined by the maximum distance the tsunami could travel in a few hours. A tsunami watch with additional predicted tsunami arrival times is issued for a geographic area defined by the distance the tsunami could travel in a subsequent time period.

Under extreme conditions on coastlines where tsunami are recurrent, for example, the Sanriku coast of northeastern Honshu

Island, Japan, the Government has gone to extreme lengths to build seawalls behind beaches to protect towns against tsunami.

Relatively lesser attention was paid in the Indian ocean area. *This is* because tsunamis are rare in the Atlantic and Indian oceans compared to the Pacific Ocean. The recent Indian Ocean tsunami (December 26, 2004) resulted in devastation in several countries and death of about 2, 30, 000 people around the Indian Ocean. As compared to the most severe tsunamis that occurred over the last 30 years in the Pacific, the loss of lives in the Indian Ocean tsunami was higher by several orders. This necessitated the development of Tsunami Warning System for Indian Ocean.

- In the Indian Ocean there are two tsunamigenic source regions namely Andaman Sumatra subduction zone (Bay of Bengal) and Mekran subduction zone (Arabian sea)
- Most of the coastal areas in country are low lying and vulnerable to oceanic disasters such as tsunamis, storm surges, sea level rise.

The 24x7 Indian Tsunami Early Warning Centre at INCOIS, Hyderabad has the responsibility *to* continuously monitor, detect tsunamis and issue advisories to the coastal regions. This centre also acts as one of the Regional tsunami advisory service provider for the Indian Ocean region.

It may be mentioned that tsunami warning systems have some limitations. That is why there are some times false alarms or else the warning has to be withdrawn after some time. The tsunami due to very great Japan earthquake of 11 March 2011 took only about 10 to 15 minutes to reach the coast which was insufficient to warn the people. In such case, urban planning based on risk and vulnerability maps have better potential to mitigate disaster. Also, deep sea internet is in the process of development which would connect data buoy networks that detect tsunami and then transmit to laptops, smartphones and other wireless devices in real time.

Observational network of Indian tsunami system

This network makes use of data transmission through satellite for quick processing.

- Network of 17 Indian broadband seismic stations for near real time earthquake detection and real time data access to more than 300 Seismic stations scattered around the world.
- Network of 7 Indian Tsunami Buoys (bottom pressure recorders) for detection of tsunami waves in the deep ocean and access to data from the tsunami buoys deployed by other countries in the Indian Ocean and elsewhere.
- Network of 36 automatic Indian tide gauges and real time data access to more than 70 tide gauges along the Indian Ocean coast installed by other countries to confirm and monitor the progress of Tsunami waves.

The data from surface buoys in Bay of Bengal and Arabian Sea is sent via Iridium satellite to ground computers at the Tsunami Warning Centre, INCOIS, Hyderabad. This centre monitors all related data flow in real time and then issues warning.

False alarm due to tsunami

(i) Errors in epicentral parameters namely location, focal depth and magnitude of earthquake,

(ii) Complex mechanism of earthquakes

As mentioned earlier, the number of earthquakes in oceanic areas is very large but the tsunamigenic earthquakes are only a few. It may also be noted that after the occurrence of Sumatra earthquake (Magnitude 9.3) in December 2004, another great earthquake (Magnitude 8.5) occurred about 150 km south-east of the first earthquake on 28 March 2005. However, it gave rise to only a minor tsunami near Indonesia. Also, the great earthquake on April 12, 2012 near Indonesian coast did not cause tsunami as the associated fault movement was strike slip type. The tsunami warning was issued soon after the occurrence of these earthquakes near the Indonesia coast, but due to absence of signals in tsunami buoys, the warning was later withdrawn.

Dissemination of warning is done by GTS, SMS, Fax, E mail, IVRS, multi lingual web site, video conference facility, electronic display boards.

Early warning strategy for earthquakes

Earthquake hazard assessment of a region is based on the past earthquakes, tectonic models and earthquake monitoring. In this context, India has been divided into four seismic zones and building codes have been prepared accordingly. These are used as long term measures for protection against earthquake hazard. However, the public is interested in short term prediction of earthquake. At the present time, however, there is no short-term earthquake prediction capability with any degree of reliability. The processes that give rise to earthquake hazards are fundamentally complex in nature due to the heterogeneous nature of crustal rocks and the uncertain state of the forces that trigger the events. Prediction of effects after an earthquake but before the arrival of seismic waves, may help to reduce earthquake impacts. High-speed computational and microwave communication systems have made possible another aspect of prediction that is rapidly developing, namely, to predict effects after an earthquake has taken place, or while the hazard is ongoing, but before the damage has occurred. It is possible to detect the occurrence of an earthquake and send warnings before the seismic waves reach places more than several tens-of-kilometers away using microwave communication to sound alarms or take other measures. Early warnings can enable individuals in vulnerable situations to protect themselves or others. School children could take cover to avoid injury from falling structural debris or nonstructural buildings. Surgeons can suspend delicate operations. Businesses and industries can stop critical processes such as the handling of toxic substances and protect assets such as active databases. Utilities and transportation lifelines can take preventive action to avoid major service disruptions.

The basis of earthquake early warning systems (EWS) is that the greater the distance from the epicenter, the more is time for warning.The seismic waves generated at the earthquake focus propagate outward at velocities of about 3-7 km/sec, and the more destructive waves travel at even lower velocity. Therefore, it would take the first wave at least fifteen seconds to reach a distance of 100km from the focus, by which time the parameters of the event can be determined and disseminated automatically by modern communication systems using V-SAT technology.

CHAPTER 12

Microwaves in volcano eruption warning

Microwave communication has enabled to telemeter data via satellites. Synthetic aperture radar and satellite thermal anomaly sensors monitor deformation and temperature data over volcanic regions. Global positioning systems supplement data about crustal deformation in these regions.

Volcanic eruptions are the most violent and spectacular manifestations of nature. They have, therefore, fascinated man for appreciation of beauty and many times fear.People have to be evacuated due to surging lava, heat or explosion close to eruption site.

A volcano is defined as a place where molten rock and or gas is forced from the interior of the earth to its surface causing violent explosion due to liquid magma or lava flow. A volcano consists of a fissure or cylindrically shaped opening (vent or pipe) which reaches from the surface to a source of magma far below.

Volcano characteristics

Volcanoes have conical shapes and consist of layers of volcanic material and are marked by craters in the cap of some volcanoes. A caldera forms due to collapse of the tops. The causes of volcanic disaster are attributed to bombs, pyroclastic surge, lahars (volcanic mudslides) and lava low.

Geographical distribution

Volcanoes and earthquakes generally occur in the regions where two plates collide. All around the Pacific Ocean, earthquakes and volcanoes are found which is commonly called the ring of fire. Volcanoes generally erupt in the following places:

i. Aleutian Island
ii. Kamchatka Island
iii. Kurile Island
iv. Japan
v. Indonesia
vi. New Guinea

The only active volcano in the Indian sub continental lies in the Andaman sea about 135 km away from the Port Blair towards east northeast. It erupted in 1991 after nearly 200 years. It coincided with that of Mt. Unzen in Japan and Mt. Pinatubo in Philippines. Other volcanic Island in the region is the extinct volcano of Narcondum Islands which lies about 135 km north northwest of Barren Island.

Causes of volcanoes

As mentioned earlier, the cause of volcanoes is based on plate tectonics. They may form in three ways:

i. Hot spot volcanoes appear where magma from the mantle melts, causes holes and erupts to the surface.
ii. In the rift regions where the two tectonic plates separate from each other due to the force of rising magma.
iii. Where two tectonic plates collide, one plate may slip below the other, causing surface rocks to melt. The magma rises through the weak spots in the over lying plate.

It is estimated that there are about 1330 active volcanoes in the last 10, 000 years in different parts of the world. Of these,

565 volcanoes are found to be active since 1500.Typically, about 50 volcanoes are active each year.

A question arises why the volcanoes have violent nature like 'Anar' in fireworks. Volcanoes generally occur along the subduction plate (bent plate) whose rocks melt and are mixed with gases to form magma. Most of the magma appears to originate between 50 and 300 km below the earth surface. As magma rises towards the surface, the pressure becomes less and some of the gas starts to come out of liquid forming bubbles. As magma continues to rise the bubbles grow, increase in number and begin to merge with each other. Eventually, the gas pressure may become so high that the enclosed liquid and the gas is burst into the air in fountains of molten rock. These diffuse into tiny droplets and harden into fine ash. The bursting out of gas at the surface may cause chain reaction. Thus violent expansion of gas propagates to deeper levels of magma which starts rising and forming further bubbles (Fig 12.1). The eruption of volcano consists of a series of such blasts. They may last for a few seconds to several days. In short, a violent explosion should have magma rich in gas.

Fig. 12.1 Volcano eruption (NOAA)

The estimates of magma in different volcanoes are made by several geo physical methods. Many of these are routinely monitored by seismographs and other instruments. The rumbling of magma causes two different signatures on the earthquake records.Some are high frequency earthquakes and others low frequency earthquakes. In low frequency earthquake, the primary and secondary (s) waves are rarely observed and, therefore, it is difficult to locate the source of volcano. However, high frequency signals enable to find the epicenter of tremors.

Warning of volcanoes

Predictions of volcanic eruptions at Mount St. Helens reduced risk to life and property, eased concern of people living near the volcano and allowed timely evacuation. Although volcanic prediction uses many geophysical methods, the advent of microwave techniques like synthetic aperture radar and global positioning system besides rapid communication of volcanic tremor data through microwave channels enables to issue timely and reliable warnings for volcano eruptions.

Warning is generally based on the following methods:

i. Seismic method
   As mentioned earlier, monitoring earth tremors in volcanic zones is now the most common method of eruption warning. The ascent of magma causes rumbling of the earth. Thus, there is usually a marked increase of tremors of shallow focal depth prior to a volcanic eruption.
ii. Temperature measurement
   The temperature of crater lakes, hot springs often show a sharp increase before eruption. The infra-red thermal imaging in several bands in the satellites at regular intervals enable to issue warning when the temperature at the volcanic region rises.

iii. Gas composition

The composition of gases may vary before eruption.

iv. Gravity and magnetic

Movement of lava at depth cause slight changes in local gravity and geomagnetic fields.

v. Ground deformation

Before an eruption, deformation of volcanic region is caused by faulting and movement of magma and geothermal fluids in the subsurface a volcano. This swelling of earth or ground deformation may be detected by synthetic aperture radar data from satellites and aircrafts. SAR system has several advantages over ground based deformation measuring methods. Pre-existing ground control is not necessary for SAR measurements while ground based methods require base line measurements for comparison. SAR also provides continuous coverage over a wide area, where as existing methods provide data at a limited number of points on the ground .Thus SAR interferometer is capable of mapping the entire deformation region around active volcanoes(Fig 12.2).

Fig 12.2 Synthetic aperture radar interferogramaround volcano(USGS)

Quantitative observations are made by microwave radio meters, high resolution radars, measurements of Sulphur Dioxide discharge rate from satellites, laser altimeters and monitoring of submarines volcanoes by hydrophones.GPS point measurements have shown promise to provide improved determinations of interseismic strain accumulation along active faults which allow better assessment of volcano tremors potential.

The two most active volcanoes separated by only 30 km in the world are fueled by the Hawaiian hot spot. Satellite data of ground deformation and modeling suggested simultaneous inflation due to increased magma supply prior to eruption during 2006-2008. Increased carbon dioxide emission and shallow silent earthquakes also occurred as precursory signals.

NASA's Airborne Science Standard C-20A aircraft carrying synthetic aperture radar completed mission to study active volcanoes in Alaska, Aleutian and Japan islands to understand the deformation of the ground prior to, during and after an eruption. ERS-1/ERS-2 synthetic aperture radar interferometer was used to study the 1997 eruption of Okmok volcano in Alaska. It showed subsidence (sinking) about 16 minutes before the eruption and 5 minutes after the eruption. INSAR is preferred for active volcanoes because of its relatively rapid, spatially complex deformation even in remote and hazardous areas. It is difficult to use in the areas covered by snow. But future missions used L band radar to overcome many limitations.

GPS is used to monitor displacement rates (cm/year) in volcanic eruptions. It may, however, be noted that the errors of the height component are 2 to 3 times higher than the horizontal components. Hence, the information about the vertical component is not reliable. GPS receivers are set up over several volcanoes bench marks at the same time. This enables simultaneously to collect data from same satellites. The error caused by atmosphere and location of satellites becomes the same for all sites. Thus, their positions are determined relative to one another to accuracy of few centimeters.

CHAPTER 13

Microwaves in landslide and glacier monitoring

Microwave active and passive remote sensing methods are being used widely to monitor landslides and glaciers which occur over the mountainous region. The instruments deployed for this purpose are GPS, synthesis aperture radar and satellites with suitable wave lengths for respective remote sensing methods.

Landslides

Landslides are defined as the slippage of soil, rock and organic material along the slope of the mountains. They block roads, dislocate traffic sometimes lasting for several days and trap people, cars and animals on their way. The tragedy of Kedarnath in June 2013 in which about 10, 000 people were washed away or reported missing (presumed dead) was attributed to a massive landslide due to a cloud burst and heavy rain (Fig 13.1). It also caused flooding over a large area. Most of the destruction was caused by sudden and rapid melting of ice and snow on the Kedarnath mountain in Uttarakhand. Destruction of bridges and roads left about 100, 000 pilgrims and tourists trapped in the valleys. The Indian Air force, Army, and para military troops evacuated more than one lakh people from the flood rampaged area. After receding of the floods, satellite images showed one new stream near Kedarnath town. National Disaster Response Force,

Public Works Department and local administration extended help in the rescue operations. Helicopters were also used to rescue people but difficulties were faced because of rough terrain, heavy rain and fog.

Fig. 13.1 Landslide in Kedarnath, 2013

Landslides occur time and again during rainy season all along the weak zones in the Himalaya, in some areas of the Sahayadri range along the west coast and in Nilgiri hills in south India. Massive landslides have also been reported from time to time during earthquakes in different places in the world. The causes, speeds and destructiveness of landslides vary widely. Some landslides occur more slowly while others can transform suddenly into mud or debris flow that can travel thousands of kilometers in very short time causing massive destruction and loss of life. Many landslides take place during or after heavy rains or snow melt when ground water pressures increase.

Causes of landslides

The major causes of landslides are as follows:

i. Excessive rainfall or rapid snow melt.
ii. Earthquakes or volcanic eruptions causing vibrations.

iii. Inherent weakness in geological settings
iv. Deforestation and soil erosion.
v. Unplanned constructions and other anthropogenic activities, mining, excavation, water leakage.

Himalayan and other mountains are formed due to geological forces called tectonic forces. Their upper parts are most fragile and vulnerable to slope failures. Due to increased construction of roads in the recent times, unplanned housing, commercial activity and cutting of trees, the number of landslides related to stability of cut slopes has increased. Landslides are more common during monsoon season from July to September in Indian region. However, the early onset of monsoon triggered landslide in Kedarnath in June 2013 causing major disaster.

Landslide monitoring

Landslide studies are organized into 3 phases namely detection, monitoring and prediction of slope failures.Instruments are used to measure rainfall, pore water pressure and sinking of ground (subsidence) settlement. Some of the equipment required for monitoring landslides are:

i. Automatic rain gauges – to measure rainfall.
ii. Piezometers – To measure water pressure.
iii. Extensometer – To measure settlement of ground
iv. Precision leveling – To measure slight movement of rocks.
v. Crack monitors
   The accuracy required for the measurement of landslide displacement should be at least a few centimeters.

Specialized observations

i. Geodetic: Global Positioning System (GPS) built around the geographic Information System (GIS).This system is used for

a. Point positioning
b. Precise surveying
c. Differential changes in positioning

The advantage of this method is that it does not require line of sight between stations. GPS also works in bad weather.This is considered to be reliable tool for land slide monitoring.

ii. Synthetic aperture radar interferometers (SAR) . This method is very useful because it is able to measure small changes in surface topography from satellites.SAR enables to detect the local instabilities which may not be detected by ground measurement. Synthetic aperture radar (SAR) fringes generally do not show rapid variations except over a small area close to a landslide. The increased accuracy can show relative motion in different parts of the landslide.

iii. Ground penetration radar (GPR).

The georadar using microwaves has high resolution surface mapping capability upto a depth of about 20 meters below earth's surface. It may help landslide monitoring along a sloping plane.

iv. Acoustic emission: Using special acoustic emission detection, it is possible to provide crucial information in advance before a land slide.

v. Micro earthquakes

The weakening of the rocks and their slight movements can also be detected as very small tremors called micro earthquakes which are recorded by seismographs. The multichannel seismic system is also used to determine seismic wave velocities which show a change when the rocks start weakening.

vi. Earth's resources satellites allow us to monitor the vegetation covers and glaciers which may also slide down the slope causing similar causalities. They are specially useful for polar studies because these regions are generally obscured by darkness (polar night) for long periods. Earth resources satellite has day and night sensing capability with active microwave instruments.

Landslide mitigation

Since landslides are caused by the instability of hill slopes, various techniques are used to increase their strength as remedial measures.

The increase in the forces which destabilize the slope are caused by

i. Addition of load on the top of the slope.
ii. Removal of material from the top of the slope.
iii. Increase in the steepness of the slope; steeper slopes are more vulnerable as compared to gradual slopes because of earth gravity.
iv. Increase in pore water pressure in the slope mass, caused by heavy rains or waterfalls.

The decrease in the strength of the soil may be due to

i. Increase in pore water pressure.
ii. Liquefaction of loose saturated granular soil.
iii. Weathering of the slope mass.

In India, several measures are taken to protect hill slopes instability. The land slide corrective measures are:

i. Direct measures to control landslide by retaining walls, rock bolting, rock anchors, soil nailing, earth embankment.

ii. Indirect methods like slopes grading surface and sub surface draining measures to allow heavy rainfall to flow and growth of vegetation.

Land slide zonation maps have been prepared for Himalayan region. These are based upon geographical information system (GIS), mapping of landslides vulnerable areas and combining different data sets. According to this method, land slide prone areas are classified as high, moderate and low.

Glaciers

Glaciers are large mass of ice that occur in areas where rate of snow precipitation is greater than melting of snow. A glacier constantly changes its shape and form to adjust with changing climatic conditions. It is believed that from 2 millions years ago to 10, 000 years BP, glaciers occupied 30% of total area of earth's land surface. Their coverage is however reducing which is attributed to global warming caused by anthropogenic activities. The glaciers occupy 20% of the world average in Himalayan range. These glaciers differ from those in other parts of the world such as Alps and other high altitude glaciers in Norway. They have high melt rate resulting into large water flow to rivers.

Glaciers influence weather and climate on different scales. The snow and ice cover over Himalayan mountain form a cold natural water reservoir. This regulates the water supply throughout the year in the rivers originating in Himalaya. Glaciers act as buffers and regulate water run off from high mountains(Fig.13.2). Fluctuations in glaciers can effect river catchments and availability of total water. Since 1986, a World Glacier Monitoring Service has been established under United Nations Environment program.

Fig. 13.2 Himalayan Glacier (The Telegraph, U.K.)

Factors effecting glaciers

i. Anthropogenic warming
ii. Changes in pressure systems in the north Atlantic or El Nino conditions in the Pacific oceans
iii. Slow changes in ocean currents.

Glaciologists measure the annual gain or loss in thickness of a glacier called mass balance in meter water equivalent. One meter of ice thickness corresponds to about 0.9 meter water equivalent.

Satellite monitoring

Glaciers are monitored through scientific expeditions. More extensive coverage is, however, possible through satellites which map the glaciers by instruments called advanced spaceborne thermal emission and reflection radio meters. Several countries of the world are collaborating in a Global Land Ice measurement program from space. The potential of using satellite radar interferometers to monitor ice flow and ground line positions of ice streams was first studied for Antarctica region . SAR data also gave ice sheet velocity over a distance of one km in the region. The World Glacier Monitoring Service collects observations on changes in mass, volume, area, glaciers fluctuations as well as glacial inventories. These are valuable for climate change monitoring. Most of the studies have suggested that glaciers are melting rapidly all over the world.The most comprehensive picture of the rapidly thinning glaciers along the coastline of Antarctic and Greenland ice sheets has been detected by satellite lasers. It has been estimated from a satellite called Gravity Recovery and Climate Experiment (GRACE) that 300 billions tons of ice is being lost every year from the Antarctic and Greenland glaciers. This satellite is able to detect tiny variations in earth's gravity resulting from changes in mass distribution besides movement of ice into the oceans which causes sea level rise. Using these changes in gravity, the state of the icc sheets can be monitored at monthly intervals. This will help to better identify and predict melting and corresponding sea level rise. Passive microwave satellite sensors have also been used to track glacier snow melt and resultant snow melt induced floods using 37 GHz brightness temperatures.

CHAPTER 14

Microwaves in warfare

The conventional methods like artillery and tanks have been used in the war since long. Chemical and biological methods have sometimes been used although it is against Geneva convention 1977. The real breakthrough in fighting the enemy has been provided by electronic warfare through land, air and sea forces.

Microwave technology was first deployed in radar during World War-II for the detection of enemy aircraft and submarines. Since then, microwave applications in weapons for air, sea and land has been expanded by using active and passive methods of remote sensing. Electronic warfare has immensely enhanced the capability of armed forces during attack, protection and support. It also uses methods to reduce the capabilities of enemy radar, missile and other equipment. During the Iraq war, high power microwaves and other electromagnetic weapons were used by U.S. military to disrupt and destroy Iraqi defence.

During war, the following weapons are used to attack the enemy:

(i) Guided bombs.
(ii) Rockets
(iii) Shells or projectile fired from a gun
(iv) Torpedoes in sea.
(v) Hand grenades and laying of mines
(vi) Guided missiles.

Of these, guided missiles (Fig.14.1) have been developed to target areas from a few hundred to thousands of kilometers away. These have the capability to carry different types of ammunitions including nuclear bombs. Major attention is, therefore, given to take measures to destroy them during the transit before they cause destruction.

Major components of missile system:

i. **Guidance systems**

Guided missiles use infra red, laser or radio waves to guide them for striking the target. Advantage is taken of the radiation or radio waves from encmy radar or the heat of an engine. Alternately, this is supported by laser beam emitted through the infantry or radar of the launch vehicle. The steering of the missile is usually done through a computer. The guidance system like GPS accordingly corrects the path of the missile till it strikes the target.

ii. **Fight System**

This is required to maneuver the missile in flight due to movement of target or other inaccuracies in the missiles.

iii. **Engine**

Missiles have jet or rocket type engine. They have sometimes multiple engine stages. Surface launched missiles have both types of engines, normally a rocket booster for launching and a jet engine for sustained flight.

iv. **War head**

The war head contains explosives to destroy target which may be nuclear weapons, chemical or radiological weapons.

The following types of missiles are generally used:

(i) Surface to surface
(ii) Surface to air
(iii) Air to air

The nuclear weapons need to be targeted more accurately to cause localized damage. Their trajectory (path) is computed by ballistics and small corrections in the flight path are applied as required. Although primarily meant to carry nuclear arsenal, ballistic missiles are also used to strike ships. They can be launched with mobile launchers, ships or even submarines. Drones (unmanned aircrafts) are also being used to fire missiles at specific targets.

Fig. 14.1 Guided missile Patriot

Some missiles for specific purpose are as follows:

i. **Cruise missiles**

   They are mainly launched from air, sea or submarines platforms. The Brahmos cruise missiles are a joint venture between India and Russia.

ii. **Anti-ship missile**

   In addition to anti-ship missile, anti-submarine missiles are also available. They use the missile to deliver another weapon system like torpedo. The second weapon undertakes underwater phase of the mission.

iii. **Anti-tank guided missiles**

These missiles may be launched from aircraft, vehicles or by ground troops.

iv. **Anti-aircraft missile**

Surface to air missiles can be launched from different types of platforms. Portable systems can also be used.

v. **Anti-ballistic missile**

These are used for defense against short range missiles. They carry explosives, war heads or destroy the target only by collision.

vi. **Air to air missile**

Radar has long been used to keep surveillance about the location of enemy aircrafts or ships. It is also used to gather information, detection, identification and warning for accurate assessment of enemy strength. Modern day electronic attack involves the manipulation of electromagnetic energy to reduce the effectiveness of the enemy radar.

Advanced version of radar called the phased array radar is used for tracking of missile. It uses a very narrow beam of microwaves. A very large antenna tens or hundreds of meter is required to produce narrow beam. During war, such a large antenna cannot be moved easily. Hence phased array canopy was developed. A phased array is a set of hundreds or thousands of little radar sets that emit microwaves with their peaks lined up in a certain direction. By changing the relative spacing of the peak electronically, the resulting narrow beam can be steered anywhere in less than a second.

High power microwave weapons system has been developed to destroy the electronic instruments of the enemy. These weapons can be sent by a missile and then exploded near the target to produce a pulse of high power microwaves. These waves damage electronic equipment used for communication and the computers. They may affect software and computer memories. The very intense pulse of microwave power can get into the electronic equipment. Power cables and telephone lines wiring can also allow microwaves to enter the equipment and cause damage.

Spy Satellite

Some countries have launched spy satellites to keep a watch on other countries. Modern warfare has been made easier by microwaves transmitted and received via satellite providing reliable communications. The communications are intercepted and decoded for spying. However, China is reported to have tested techniques to destroy low orbit satellites.

The global positioning system (GPS) receivers use microwave signals from satellites to find the position of a target almost anywhere on the earth with great precision.Modular multi-functional defensive information system (MIDIS) has been developed using latest principles of signal selection and processing with microwave modules.

i. **Directed energy Device**

   Like laser, coherent microwaves and radio frequencies are emitted through amplification by simulated emission. This device is called a maser. Common types of masers are called atomic beam masers (using ammonia or hydrogen), gas masers (rubidium based), and solid state maser (like ruby).

   Other system for this purpose is called microwave gun or cannon.High power microwave pulse system envisages to disable electronic system of enemy vehicles. Microwaves against human beings can cause a burning sensation or drowsiness. Their main advantage lies in their penetrating power through most common building material. These radiations can be focused on individual or a large area with specialized antenna. Also, they force an enemy to flee rather than fight due to a sense of anxiety caused by stray noises created by specialized antennas.

Crowds are often controlled by police and during emergency situation (like riots). They generally use gas rubber bullets, batons and water cannons to disperse the crowd.A new weapon called microwave cannon or heat ray has been developed to disperse crowds with the help of high pressure beam of high frequency microwaves of 95 GHz (3.2 mm wavelength). Its effect is similar to that of microwave oven which heats food at 2.45 GHz. It is also called Active Denial System. The dielectric properties of human body consisting of fat and water molecules produce sudden heating and intense pain when exposed to microwaves. It has no side effects like causing cancer, birth defects or fertility problems. Only in exceptional cases, it may cause second degree burns to a few persons. However, if the atmosphere has very high humidity or it is raining, its impact may be less on human body due to partial absorption of microwaves.

ii. **Defense against enemy attack**

To protect the armed forces against electronic attack by the enemy, the following methods are used:

i. Chaff: Large number of metallic thin foils are scattered over an area by aircraft or ships to conceal or disguise guided missiles. This becomes feasible because false targets are created which reflect microwave signals from the radar.

ii. Missiles, aircrafts, ships and vehicles are specially designed to absorb or reduce microwave power.

iii. Jamming technique is used to send out false and misleading signals which confuses enemy radars. Global navigation satellite system(GNSS) jammers broadcast disruptive signals in the GNSS bands. A Jammer can overpower the much weaker GNSS signals and disrupt GNSS based services over an area of several kilometers. Since GNSS are quite vulnerable, U.S Army is searching alternate methods to limit its dependence on GPS.

iv. High frequency microwaves are directed towards any projectile that is fired at an aircraft. The system uses a missile detecting and tracking system, a command and control system and a scanning array. It consists of a passive infra-red cameras. The missile launch point is determined by the command and control system. The scanning array sends microwaves which disrupt the surface to air missiles guidance system and deflects it from the aircraft. This is called Vigilant Eagle as a part of airport defense system.Bofors black out is a high powered microwave weapon to destroy many types of commercial electronic equipment far away.A pulse radar has been developed inspired by dolphins hunt which can detect hidden surveillance equipment and explosives.

India has launched a program to use the array of geostationary satellites (GSAT) to monitor missile activity in a radius of 6000 km radius. Their capability to map anything to a resolution of one meter will allow to capture slightest movements.This project is expected to install sensitive surveillance equipment on the GSATs. The signals captured from missiles will be transmitted to a central unit. This will enable to launch or intercept missiles quickly to destroy the enemy missiles.India's first military satellite GSAT-7 (Rukmini) is a geostationary communication satellite with UHF, S, Ku and C band transponders. It will monitor Indian submarines and aircrafts and keep a hawk eye over the Arabian Sea, Bay of Bengal, Indian Ocean regions. For difficult terrain like

Ladakh or Sikkim, Indian troops will be provided with suitcase based satellite communication (satcom) terminals for better connectivity using Ku band. These can be carried by vehicles, mules or soldiers to areas where line of sight communication is not possible.

India is in the process of replacing its 150 km range Prithvi ballistic missiles with the Prahar missiles which has better precision strike capabilities. It is a single stage missile and is fuelled by solid propellants. Several designs of naval weapons (submarines) for protection against enemy through sea have been designed which could travel on the surface as well as beneath water. The main features of these submarines is to switch on from floating to travel under water or vice versa by using aircraft or sea water to make it heavier to travel much deeper. Its navigation is done with the help of computers. The navy's youngest frigate has highly advanced features to reduce radar, magnetic and acoustic signatures. It is fitted with supersonic Brahmos missile system, advanced surface to air missiles, Shitil, A-190 medium range gun, electro optical 30 mm close in weapon system, anti-submarine weapons such as torpedoes and rockets and advanced electronic warfare systems, all of which are based on microwaves.

CHAPTER 15

Microwaves in space exploration

Man has always been fascinated by the Sun, Moon and stars shining in the sky. The moon has always been considered as the symbol of beauty and love and was the central theme in many verses by the poets. Its exploration by space crafts, satellites and first landing by Neil Armstrong on 29 July 1969, revealed its detailed structure. The tales about Saptarishi and polestar (Dhruv Tara) since our childhood learnt from grandparents have always aroused our interest to know more about them. Discovery of microwaves has led to unveil many mysteries of the space and universe.

Space exploration through microwaves

Microwaves, in particular of millimeter wavelength have been extensively used by astronomers. They enable us to study galaxy formation and its evolution, stellar and planetary systems, the composition of atmospheres of solar system and the cosmic back ground radiation for big bang theory of the formation of the universe.

Astronomers use optical telescope to study space objects in reduced light pollution in the night sky. However, the scope of such telescopes is limited. The signals from celestial objects are so weak that they can be masked by manmade interference. Radio telescopes using microwaves were, therefore, developed to study distant objects in the universe by recording and analyzing radio waves emitted by these objects as well as by gas, dust and high energetic particles in the space between stars. Radio telescope are, therefore, designed to prevent interference into the receiving systems to detect weak signals. By International agreement, radio frequencies divided as blocks or bands are allotted to prevent one type of station from interfering with station of another type.

The largest space observatory to probe the universe has been started in Chile under joint efforts of north America, European and Asian space agencies. It is located in a super arid desert at an altitude of 5000 meters where there is virtually no water vapour or vegetation. It is called Atacama large millimeter- sub millimeter array (ALMA) which has 66 antennas ranging from 7 to 12 meters in diameter. It can probe darkest and remotest regions of the universe, due to least interference from sources of noise. In India, the Tata Institute of Fundamental Research, Mumbai indigenously designed giant meter radio telescope near Pune, Maharashtra. It deployed multi frequency synthesis imaging system using 30 parabolic dish antennas each of 45 meter diameter along 25 km base line. The correlation of radio signals for all 435 possible pairs of antenna or interferometers over several hours enables radio images of celestial objects to be synthesized with a resolution equivalent to that attained with a single gigantic dish 25 km in diagonal. The array operates in the frequency bands centered around 50, 153, 233, 325, 610 and 1420 Mhz. Radio astronomy is the main focus at the Indian Institute of Astrophysics, Bengaluru. Its decameter wave radio telescope is located at Gauribidnaur, about 100 km north of Bengaluru.

The exploration of space objects has been accelerated by microwave remote sensing using active and passive methods. Because of the long distances involved from the earth, space crafts or satellites are often required for communication purpose between the distant planets. They are needed to send pictures, videos and scientific information captured by satellites and space shuttles back to earth using microwaves. Their navigation, therefore, depends on radio transmissions between a distant space craft and a network of ground stations on earth. However, since the distances involved are very large, it takes hours or days to send instructions . This time delay sometimes affects a space craft to maneuver it rapidly. The ground station using astronomical telescopes use very large antenna system to receive signals from the space objects. Synthetic aperture radar has been extensively used in radio astronomy to simulate a large radio telescope by communicating observations taken from several locations using a mobile antenna.

The solar system

The solar system is a part of the milky way galaxy which contains billions of stars and clouds of dust and gas. Millions of such galaxies make the universe .

The composition of sun, its surface temperature of 6000^{0}C, sun spots every 11 years were known to astronomers since long time. The solar system consists of the Sun, eight planets (fig 15.1), satellites and other celestial bodies known as asteroids and meteoroids (Table 15.1). Sun gives us light and heat and is one of the stars closest to earth. The eight planets which revolve around the sun in fixed orbit (path) are called Mercury, Venus, Earth, Mars, Jupiter, Saturn, Uranus and Neptune. Pluto, Aries and Cyrus are dwarf planets. Only earth has favourable conditions to support life but search is being carried out to find the existence of life elsewhere in the universe. Our earth has only one satellite namely moon. It takes about 27 days to complete one revolution around the earth. The number of moons around Mars, Jupiter,

Saturn, Uranus and Neptune are 8, 16, 18, 17 and 8 respectively. New smallest moon around Neptune has been recently discovered by Hubble space telescope. The first four planets are made up of rocks while the remaining planets consist of gases and liquids. Mars, Jupiter, Saturn, Uranus have rings around them which are actually small debris.

Table 15.1: Space terminology

S.no.	Objects in sky	Definition
1	Celestial bodies	The sun, moon and all objects shining in the night sky.
2.	Stars	Celestial bodies made of gases and emit heat and light just like sun.
3.	Constellation	Group of stars form various patterns like Saptrishi (7 stars)
4.	Planets	They do not have their heat and light . They are lit by the light of stars.
5.	Satellite	A celestial body which moves around the planets.
6.	Asteroids	Part of a planet which exploded many years back.
7.	Meteoroids	Small pieces of rock which move around the sun. Sometimes they fall on the earth if not burnt in the sky.
8.	Galaxy	Huge system consisting of billions of stars, clouds of dust and gases. Milky way or Akash Ganga is one such galaxy.
9.	Universe	It consists of millions of galaxies and stars.

Indian astronomers make use of nine planets (Navgraha) to prepare horoscopes.

Table 15.2: Planets in solar system

S.No.	Planet	Diameter Km	Distance Km (10 millions)	Gravitation Earth=1	Density Water=1	Time of rotation
1.	Mercury	4850	5.79	0.26	3.8	59 days
2.	Venus	12228	10.82	0.90	4.9	?
3.	Earth	12756	14.95	1.00	5.5	1 day
4.	Mars	6788	22.77	0.37	4.1	24h 37m
5.	Jupiter	1, 42, 980	77.77	2.64	1.3	9 h50m
6	Saturn	1, 16, 000	142.56	1.13	0.7	10h 47m
7.	Uranus	51, 000	286.85	0.84	1.5	10h42m
8.	Neptune	45000	450.00	1.14	2.2	15h 48 m
	Dwarf	Planets				
1	Pluto	2330	595.00	-	-	153h
2	Aries	3400	1500.00			-
3	Ceris	933	41.90			9.075h

Light waves travel at the speed of $3x10^{10}$meters/second. One light year means $9.46x10^{12}$ kilometer distance.

Moon (Early missions)

Moon has been explored by spaceships as well as landing by man. Its internal structure through Apollo missions was obtained by keeping 4 seismometers in Alsep region. Moon quakes are caused by the impact of meteorites and are small in intensity. The composition of moon was also found from the rocks brought by the astronauts after analysis in laboratories in different countries.

The internal structure of moon was found as follows:

i. 1 to 20 Km thick top layer is made up of basaltic rocks.
ii. 20 to 60 km thick second layer made of stronger rocks, with different types of basaltic rock.
iii. 60 to 1000 km layer is called mantle.
iv. 1000 to 1740 km layer is slightly softer called the asthenosphere. It contains molten elements possibly made of iron- iron sulphide.

The large craters on its surface are attributed to the impact of meteors. Its rocks are $3.16x10^{12}$ years old and are becoming colder day by day.Due to lesser gravitational attraction, there is no atmosphere on moon. Sound cannot travel from one place to another unless special instruments are used between two persons who want to talk. Due to one sixth of earth's gravity, the weight of a person on moon reduces in the same proportion.

Recent missions.

Radio waves transmitted from earth take 1.28 second to reach moon surface.

i. **India**

The first moon vehicle called Chandrayan-1 was launched by India in October 2008. This space craft was designed using an India made launch vehicle called (PSLV-XL). Its communication system consisted of 0.7 meter diameter parabolic antenna for pay load data transmission. The telemetry and command communications operated in S band frequency. The space craft was mainly powered by its solar energy. Some power (750 W) was stored in lithium batteries for use during eclipses. It carried five Indian instruments and six foreign instruments . They were used for terrain mapping and its topography, mineral mapping, its composition and other radioactive elements in its radiation environment, composition of lunar atmosphere . The moon impact probe developed by Indian remote sensing agency consisted of a C-band radar altimeter for the measurement of the altitude of the probe. This probe successfully crash landed at the lunar south pole on 14 November, 2008. It carried a picture of India making us the fourth nation to place a flag on moon. The outstanding discovery was hydrogen bearing compounds, presumably water-ice distribution around north and south poles of moon based on Lunar Prospector data. Mini synthetic aperture radar image also showed crater Erlanger, just south of the crater Peary in north pole area(Fig 15.2). Other sensors like X-ray spectrometer detected titanium besides confirming calcium, magnesium, aluminum and iron on the lunar surface. Chandrayan's moon mineralogy mapper detected water on the moon which is considered as a major discovery.

After successful launching and operationalization of Chandrayan mission, Indian Space Research Organization (ISRO) launched Chandrayan 2 on 22 July 2019. Its lander and rover were scheduled to land in the south polar region on 6 September 2019 and conduct scientific experiments. However, its lander crashed while attempting to land due to software glitch. (Fig15.1)

Fig. 15.1 Chandrayaan 1 and 2

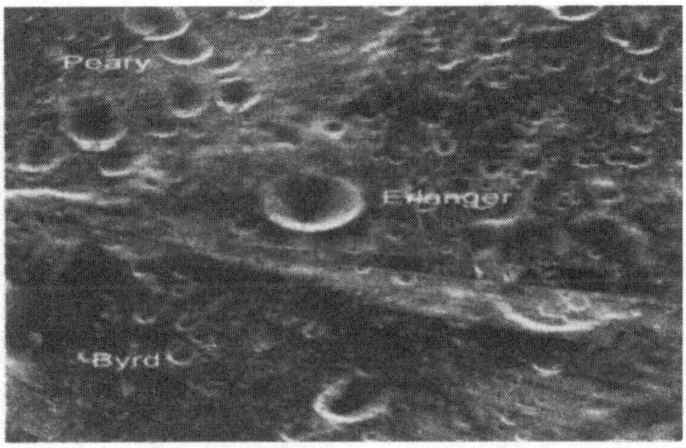

Fig. 15.2 Mini SAR observations of moon

ii. **Japan**

The first Japanese Lunar mission called Hitan launched in 1990 spacecraft failed but later in 2007, they launched the SELENE spacecraft to obtain scientific data about lunar origin and evolution and to develop the technology for the future lunar exploration. This also carried a low frequency radar.

iii. **USA**

NASA sent its Clementine mission in 1997 with the objective to test sensors and spacecraft components to the space environment and to make scientific observations of the moon and the near-earth asteroid 1620 Geographos. Later, the Lunar Prospector was sent in 1998 which was designed for a low polar orbit investigation of the Moon. It included mapping of surface composition and possible polar ice deposits, as well as the measurement of magnetism and gravity. The most important mission of the project was to detect the presence of water on moon. For this purpose, mini synthetic aperture radar designed by NASA was used to search for lunar polar ice. The instrument transmitted microwaves of frequency 2.5 GHz. Since ice has larger reflectivity, the microwave sensor helped in the water content estimation in moon's polar regions. The instrument discovered more than 40 permanently darkened craters near the moon's north pole. They are assumed to contain an estimated 600 million metric tons of water ice.

iv. **China**

It sent a microwave radio meter for lunar exploration in 2007.

Titan

Titan is the moon of planet Saturn. Its atmosphere consists of 90% nitrogen and rest methane. Its surface temperature is 180^0C below zero. It has been explored by many countries. USA sent several missions called Pioneer in 1980 and Voyager 1 and 2 in 1981. The European Space Agency also sent space mission to Titan in collaboration with NASA in 2004 by the Cassini – Huygens space craft. It mapped Titan's surface by radar. It found lakes of hydrocarbons in north polar region.

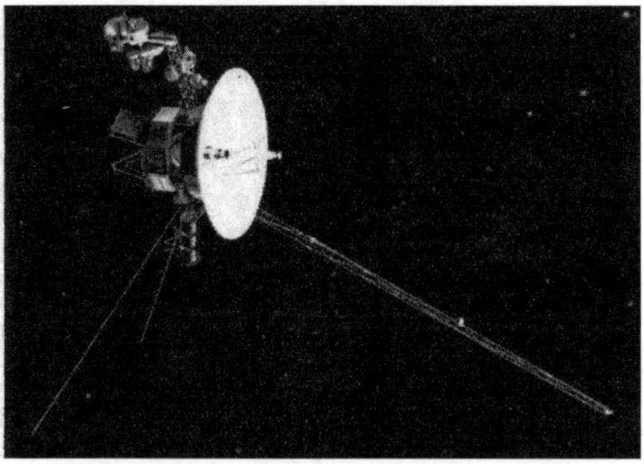

Fig. 15.3 Spacecraft

Mars

Mars is called the red planet which is attributed to iron oxide present in its sandy soil. Its atmosphere contains carbon dioxide

(95%) and nitrogen (3%) with negligible oxygen. Dust raising winds blow on its surface. Its average temperature is -55 ^{0}C. It has huge craters, volcanic ridges, long and wide deep valleys. Earlier attempts launched called Mariners 4 in 1965, Mariners 9 in 1971 by USA and Mars Fobos (1988) by Russia gave lot of information about the planet. But the landing of Viking -1 on its surface gave valuable data about its surface and climate. In 2004, two robot vehicles launched by NASA landed on the surface of Mars . It indicated large flow of water sometimes in the past.

Different missions were undertaken for study of Mars like Viking and Mars Odyssey 2001, Mars Express 2003, ESA ad Mars Reconnaissance Orbiter, 2005 . Under NASA Mars exploration program, a miniature microwave soil moisture sensor capable of providing shallow sub surface of Mars was designed. Three different frequencies were used to measure dielectric constant and loss which depend upon the sand and soil in the frequency range 0.6 to 10 GHz. The methodology enabled detection of brines at different depths.

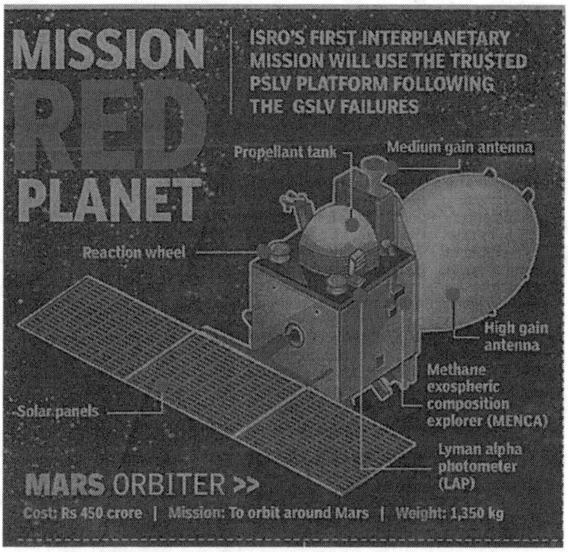

Fig 15.4 Mars mission (India)

Mars Science Laboratory is a space mission to Mars launched by NASA on 26 November, 2011. It successfully launched a Mars rover called Curiosity to accurately map in a small target landing area called Gate crater on 6 August 2012. The objectives of this mission were to study its climate, geology and surface radiation, Mars habitability and for collection of required data or a future manned mission. The rover carried many scientific instruments. It has two computers with 256 KB, 256 MB of DRAM and 256 MB of flash memory. Command signals are sent to the rover by both X-band and UHF bands. The X-band small deep space transponder communicates directly to Earth. It communicates with two Mars satellites called Mars Reconnaissance orbiter and Odyssey orbiter. The orbiters have large antennas and more powerful radars. Radar data taken from orbit revealed complex morphology of buried channels in young regions of Mars. Curiosity sample analysis at Mars suggest that red planet lost its oxygen four billion years ago after massive collision with a large object. The Mars mission by India is shown in Fig 15.4.

The issues of using inter planetary internet arise due to slow speed of light for such long distances. A one way radio signal from earth to Mars takes between three and half hours and 20 minutes which becomes twice for round trip time. The other problem is planetary rotation. This results in break in the communication as Mars surface changes its position, The solution is not easy at present. NASA 2020 rover will play a key role for a human mission to Mars.

Venus

The planet Venus is always covered with clouds. Satellites like Mariner 1 and 2 were sent by USA to explore Venus. The Japanese satellite, Akatsuhi was used to image in ultraviolet, infra red, microwaves and radio regions. The studies included lightning in clouds and volcanism on the planet. The Vanera 15, Vanera 16 and Magellan space probe mapped the surface of Venus using synthetic aperature radar.

Jupiter

Information about Venus was obtained by space crafts Pioneer 10 (1973) and Pioneer 11 (1974). The atmosphere over Jupiter is thousands of kilometer thick consisting of hydrogen, ammonia and methane gas. The planet is characterized by several red coloured stripes. The great red spot has been found in southern hemisphere which is several thousand kilometer long. The Judo aircraft launched in 2011 was aimed at studies of the atmosphere of Jupiter using microwave radiometers. It may be mentioned that the most distant space craft launched by NASA in 1977 in the universe is the Voyager 1. It was travelling at a speed of 3.6 AU per year. One Au means the distance between the sun and earth is 93 million miles. Since Voyager 1 was 9.1 million miles from the sun, it takes about 13 hours and 39 minutes for microwave signals to reach the space craft from the ground station of NASA. Alternately same time is taken for the signals from Voyager to ground antenna. The space craft Galileo also sent lot of information about the planet. Its temperature remains 180°C below zero. Its atmosphere has 90% nitrogen and rest methane.

Saturn

Space borne imaging radar missions carrying synthetic aperture radar on the space shuttle were used for Saturn. Cassini was used to map its surface. This space craft showed a monster hurricane at the Saturn north pole where its eye was 20 times more as compared to that on earth. It is interesting to note that the winds blowing around the eye wall of the hurricane are four times stronger than those over the tropical regions. The later type of tropical cyclones get energy from water vapor, but the cause of Saturn storms is not understood.

Mercury

Mariner 10 carried out detailed structure of Mercury mapping 40 to 45% of its surface. However the space craft Messenger launched on August 3, 2004 gave information about the presence of minerals on its surface, height of land formed, its magnetism and atmosphere. Its probe discovered large amount of water in the planet. It also provided evidence of post volcanic activity on its surface and gave an evidence of a liquid planetary core.

NASA's Messenger space craft in orbit around Mercury took photograph of earth recently. It is expected to throw light on the composition of Mercury and its atmosphere.

The space probe programme of USA using X band microwave communications are

- Viking Mars landers.
- Voyager mission to Jupiter, Saturn and beyond
- Galilieo Jupiter Orbiter
- New horizons mission to Pluto and Kuper belt
- Cassini – Hugens Saturn Orbiter.

Sun

Although major layers in sun have been investigated through several space probes, detailed information about the layer sandwiched between the solar surface and the corona was lacking. This is the layer through which the energy necessary to heat the solar outer atmosphere is transmitted. Also, the magnetic field is restructured associated with solar storms in this region.NASA Had, therefore, launched in June 2013 the latest sun observing satellite called IRIS (Interface Region Imaging Spectrograph) to unravel the mysteries of the Sun. It will throw new light on the corneal heating puzzle and the mechanism of largest storms in the solar system.

Stars

Molecular clouds collapse to form stars and planets provided they are able to radiate away the thermal energy generated by gravitational contraction. In the year 1998, NASA launched a sub millimeter wave astronomy satellite (SWAS). It was deployed to study the star formation by measurement of molecular cloud composition and chemistry. The cooling mechanism facilitates cloud collapse and large scale structure of the UV illuminated cloud surfaces. Its operative frequencies in the two receivers were about 490 and 550 GHz with a band width of 1.4 GHz which carries broad band (700 MHz) detectors.

The installations on this site consisted of

i. Antenna
ii. Two heterodyne receivers
iii. An acoustic optical spectrometer.
iv. The thermal control system
v. Instrument control electronics
vi. The star tracker
vii. The space surface portion of the satellite comprised of the following.

Microwave cosmic back ground radiation (CMBR)

Big Bang theory is considered as the model to explain what happened at the very beginning of our Universe. Edwin Hobble in 1929 realized that the entire universe was contained in a single point in space. The Universe was born in this event called the Big Bang. The cooling of the universe and its expansion started thereafter. The afterglow of the Big Bang is still observed and is called Cosmic Back Ground Radiation (CMBR). It is a faint glow in the universe and is considered a finger print of the Big Bang.

CMBR Mission

i. Cosmic back ground explorer (1972) was launched to study the phenomena. They mapped hot and cold spots in cosmic back ground radiation . These spots were related to the gravitational force in the early universe. They laid the foundation of giant clusters of galaxies in the universe.
ii. Welkinson microwave anisotropy probe surveyed the entire sky measuring temperature difference of the microwave radiation that is nearly uniformly distributed across the universe. It showed pictures of hot and cold regions.
iii. Plank (European space agency) was launched in 2009. It made most accurate maps of the background radiations.

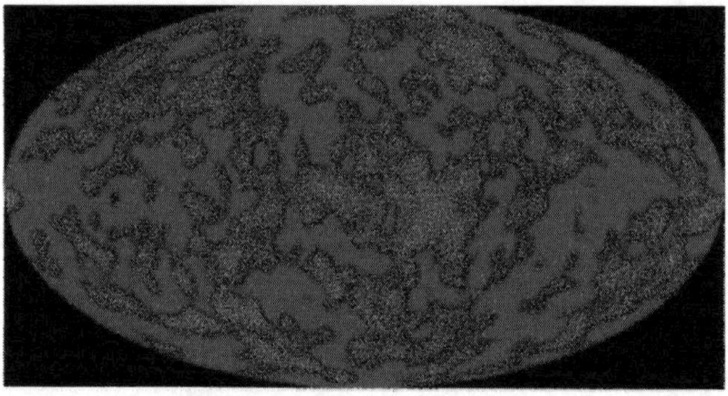

Fig. 15.5 Microwave cosmic back ground radiation

A black body is defined as an ideal substance that absorbs all the radiant energy incident on it. It emits radiant energy at the maximum possible rate per unit area at each wavelength for any given temperature.

The microwave radiation from stars and galaxy in outer space is measured to understand their formation. This radiation cannot be observed in space from the ground due to enormous distances. Therefore space based telescopes are used. A number of space probes have been launched to map the entire sky in the microwave band with higher resolution.

The stars appear to be blue in colour with naked eye. If we examine them with an optical telescope, the space between stars and galaxies in the back ground will appear completely dark. If we examine it by a sensitive radio telescope, a faint back ground glow is seen in all directions. This is not associated with any star, galaxy or other object. If this noise was associated with any object down below like the earth, it would appear to come from only one direction but the glow occurs in all directions. In the radio spectrum, this glow is strongest in the microwave region. Thus microwaves allow us to peep into the universe at its birth. In terms of thermal black body temperature, it corresponds to a temperature of 2.725^0 Kelvin. This is maximum at the microwave range frequency of 160.2 GHz according to Planck's law of emitted radiation. The Cosmic Background Explorer Satellite (COBE) launched in 1969 also detected for the first time slight irregularities in the microwave background radiation (Fig 15.5). The observed 200 micro Kelvin fluctuations are the seeds on which the clusters and galaxies grew. The microwave Anisotropy probe (WMAP) helped to determine the age of the universe within 10% accuracy i.e. 13.7 billion years. It also gave some idea about the universe less than 400, 000 years after the Big Bang.

Astronomers using the Atacama large millimeter sub millimeter array discovered recently the snow line in an infant solar system 175 light years from the earth. NASA is working on a project called Transiting Exoplanets Survey Satellite. It will use an array of wide field cameras to discover Transiting exoplanets ranging from earth sized planets to gas giants in orbit around the brightest stars in the neighborhood of Sun.

CHAPTER 16

Microwave hazards and mitigation

We have seen in the earlier chapters how microwaves which were once used only in radars for clouds and aircraft detection have revolutionized communication, navigation, weather forecasting, early warning system for earthquakes, space exploration, warfare and medical science. However, their usage in microwave ovens and cell phones has raised questions about the exposure of radiations and their effects on human body. A few persons are reported to have felt pain or buzzing sound in ears while a few others complained loss of hearing when using mobile phones for a long time. Many such problems can be addressed easily by avoiding over-use of mobile phones. The location of microwave towers in residential areas has also been debated. Isolated cases of cancer have been reported near the location of mobile tower on the roof of a house. Such observations, however, need detailed study. Keeping in view the exposure of public to microwave radiations, the Department of Telecommunications, Government of India prepared new guidelines for the microwaves towers and Mobile handsets.

Microwave oven

Microwaves emitted by magnetrons inside the ovens remain confined within its chamber. The United States Food and Drug Administration Centers for devices and radiological health places the limit of the amount of microwaves leaked from the oven during its life time to 2 to 5 Milliwatts/cm^2 of microwave radiation per square centimeter at approximately 5 cm from the surface of the oven. This is so small that it is not considered harmful to human beings. Microwave burns may however, be caused if our body parts get exposed resulting from malfunctioning of ovens.In very old microwave ovens leakage of radiation may occur. Although it is not fully authenticated, but the health problems attributed to such leakage could be:

i. Cataract in eye
ii. Dizziness and headache
iii. Blood disorders
iv. Child Birth defects (in women)
v. Skin cancer
vi. Temporary sterility in men

It is well known that any form of cooking damages a few nutrients. Microwaves may also cause internal damage to the delicate molecular structure of phyto nutrients minerals. Microwave ovens may cause a loss of about 30 to 40% of vitamin B_{12}. On the other hand, spinach retains nearly all its folate in microwave ovens as compared to about 77% loss when cooked on stove or gas. Thus more nutrients are generally retained in microwave cooking. Harmful effects on composition of food contents have not been substantiated so far.

While microwave ovens are extensively used worldwide, Russia prohibited its use in 1977, but the ban was later lifted towards 90's. The precaution which needs to be taken is to discard microwave ovens when they become 5 to 7 years old to safeguard against any leakage through the door because of poor sealing or wrapped door.

If we compare the mode of cooling between the electrical ovens used for browning and baking with the microwave oven, the important difference is that the former produces heat with infra red radiation. Since Infra red photons are more energetic, they may interfere or even cause chemical reactions. The pizza prepared in an oven undergoes several chemical treatments during infrared heating. On the other hand, microwave ovens heat food by vibration of molecules of its constituents which cause lesser chemical changes. It has, however, been reported that a chemical called perfluoroitanoic acid (PFOA) has been found in microwave popcorn bags, and also in non stick cookware and food wrappers. This substance has been linked to cardiovascular disease, chronic kidney problems or high cholesterol. The best method is therefore, to restrict their usage to minimum. Some magnetrons which generate microwaves have insulators like harmful beryllium oxide. Broken ceramic insulators or magnetron should not be touched. If the microwave oven is physically damaged, it should be disposed off properly. Sparking may occur if cramped aluminum foil or metal wire in food containers are kept inside the oven. Also, chemical hazards may occur near the magnetron because of high voltage supply if they are kept close to oven.

Major precautions for microwave oven suggested by reputed companies are as follows:

a. During installation
   i. The oven should be installed so that about 7.5 cm or 3 inch space is left between the oven and the adjacent walls. One side should be open.
   ii. The oven should be kept as far as possible from radio and T.V.
   iii. The oven should not be used by children and persons with reduced mental or physical capabilities or lack of experience or knowledge.It should not be used near water or near a kitchen sink or near a swimming pool.

b. During usage
   i. Use only heat resistant glass and ceramics for microwave grilling, convection or combined cooking.
   ii. Do not over heat liquids like water, coffee and tea to avoid injury to persons. This is because these liquids can be overheated beyond the boiling point without appearing to be boiling. This could result in spilling of very hot liquids when the container is disturbed during its removal from the oven or when a spoon or other utensil is inserted into the liquid. In order to reduce the risk, the container should remain in the oven for a short time before its removal. Over cooking food may also cause fire hazards.
   iii. The contents of baby food jars or bottles should be stirred or shaken to get tolerable temperature before feeding it to a baby.
   iv. Do not deep fry in microwave oven.
   v. Remember to pierce holes in eggs and potatoes while boiling in the microwave oven to avoid bursting.

Mobile phone hazards

The antenna and microwave transceivers in a mobile phone act as a wave guide to concentrate microwave radiation near antenna. There is, therefore, apprehension that it may harm human beings whose exposure may lead to cataract and brain diseases. World Health Organization reported that mobile phone may have a long term health risk which may produce carcinogenic hazard. Heavy users (30 minutes per day for 10 years) may show increased risk of brain cancer. This has, however, been questioned because the cases of brain cancer have not increased in proportion to the number of mobile phones.The

Norwegian expert committee did not find evidence that the low level radiations around mobiles and other transmitters increase the rate of cancer, impair male fertility or cause reproduction damage. However, a mobile phone may spread infectious disease by its frequent contact with hands. Also it is safer if the children and pregnant woman do not use mobile phones for a long time. People using pacemaker should not keep mobile phones in the pockets of shirts and keep them at last 30cm away from the implant.A new smart phone case made of same material as NASA space craft claims to reduce cell phone radiation and cancer risk by cutting the exposure upto 95%.

Some isolated cases of bursting of batteries of mobiles phones have been reported causing accidents. A mobile phone is reported to have burst when a person was holding it in his mouth to use it as a torch causing serious injury.

According to the panel constituted by the Ministry of Environment and Forests, Government of India in 2010, radiation from mobile towers may have some adverse effect on wild life particularly birds and bees.

Microwave towers emit large microwave radiation and therefore, the location of existing towers in schools, hospitals, and houses has been directed to be changed to an open place by the Supreme Court of India. The Department of Telecommunicates (DOT) has issued the following guide lines from 1 September 2012 for

EMF radiation norms

(a) The EMF exposure limit (Base station emissions) is lowered to 1/10th of the existing ICNIRP exposure level effective from 1 Sept. 2012. Indian standards are now 10 times more stringent as compared to 90% countries in the world.

i. Telecom Enforcement Resource & Monitoring (TERM) cells have been entrusted with the job of conducting audit on the self certification furnished by the service providers. TERM cell shall carry out test audit of 10% of the BTS site on random basis and on all cases where there is a public complaint.
ii. TEC has revised the test procedure for measurement of EMF for verification of EMF compliance for BTS towers in accordance with new standards effective from 1 Sept. 2012.
iii. For non compliance of EMF standards, a penalty of Rs 5 lakhs is liable to be levied to service provider.

(b) **Mobile handsets**

i. All the new design of mobile handsets shall comply with the specific absorption rate(SAR) values of 1.6 W/kg averaged over 1 gram of human tissue effective from 1 Sept. 2012.
ii. The mobile handsets with existing design which are compliant with 2.0 W/kg averaged over 10 gram of human tissue, continue to co-exist up to 31 August 2013. From 1 Sept. 2013, only the mobile handsets with revised SAR value of 1.6 W/kg have been permitted to be manufactured or imported in India.
iii. SAR value information is to be displayed on the mobile handsets like IMEI (International Mobile Equipment Identity) display. The information on SAR values is to be made available to the consumer at the point of sale.
iv. Mobile hand set manufactured and sold in India or imported from other countries shall be checked on random basis for compliance of SAR limit at TEC SAR laboratory.
v. The manufacturers in India shall provide self declaration of SAR value of the handset.

vi. Suitable amendments in the Indian Telegraph Rule under Indian Telegraph Act 1985 are being enacted in support of ensuring compliance of new SAR values for handsets.

vii. Manufacturer's mobile handset booklet shall contain safety precautions.

viii. All cell phone handsets sold in the market in India shall comply with relevant standards and shall be available in hand free mode. SAR values are to be mentioned during MRI scans.

The Department of Telecommunications(DOT), Government of India has also advised to take precautionary measures while using a mobile handset as follows

i. Keep distance.Hold the cell phone away from body to the extent possible.

ii. Use a headset (wired or bluetooth) to keep the handset away from your head.

iii. Do not press the phone handset against your head. Radio frequency (RF) energy is inversely proportional to the square of the distance from the source; being very close increases energy absorption much more.

iv. Limit the length of mobile calls.

v. Use text as compared to voice whenever possible .

vi. If the radio signal is weak, a mobile phone will increase its transmission power. Find a strong signal and avoid movement – use your phone where reception is good.

vii. Let the cell connect before putting the handset on your ear or start speaking and listening. A mobile phone first makes the communication at higher power and then reduces power to an adequate level. More power is radiated during call connecting time.

viii. If you have a choice, use a landline (wired) phone, not a mobile phone.
ix. When your phone is ON, don't carry it in chest/breast or pants pocket. When a mobile phone is ON, it automatically transmits at high power every one or two minutes to check the network.
x. Reduce mobile phone use by children as a younger person will likely have a longer lifetime exposure to radiation from cell phones.
xi. People having active medical implants should preferably keep the cell phone at least 15 cm away from the implant. In brief, addiction to mobile phones should be discouraged.

The following information has also been provided by DOT:

i. Headaches are not related to mobile use and there is no scientific evidence.
ii. The RF radiation is increased by mobile phones when used in a car to overcome the shielding.
iii. There is no scientific evidence that mobile phone can cause brain cancer.
iv. It is the antenna from which we should keep distance, not from tower and that too if we are positioned facing antenna at comparable height. At the ground level, the intensity of RF radiation from base station is much less.
v. The World Health Organization, many national and international organizations and independent expert groups are coordinating to investigate health effects of RF radiation.

Microwave weapons

Microwaves which cause pain or uneasiness in enemy forces have raised a suspicion that these may also be used against a person in jail or misused against innocent people. Since microwaves penetrate walls, buildings without being suspected, their use as microwave guns needs to be banned except during war. People who work on air craft and carrier docks should wear special suits to reflect microwaves from powerful radars in modern military planes.

Radar hazards

The cases of microwave leakage from radars are negligible. However, no attempt should be made to stand before the radar antenna which sends and receives microwaves to or from a target. Also hands or any body part should not be exposed to the antenna mounted in our houses to view the television programs although signals are not strong to cause burns.

Television hazards

Television screen also emits radiation which is harmful to eyes. Children may be advised in particular to see television for a short time otherwise their eyes may be damaged.

Disaster mitigation

During natural disasters like earthquakes or tropical cyclones, telephone lines go out of order. Mobile towers are also damaged making it impossible to call from a mobile phone. It is therefore, recommended that microwave towers should be designed to follow the guidelines of Bureau of Indian Standards for earthquake resistant structures and strong winds so that they are operational during natural calamities. Meanwhile the search and rescue satellite based system operating at 406 MHz may be useful to locate people in distress or shipping vessels(Fig.16.1). A number of mobile towers also need to be deployed by National Disaster Management Authority in the affected area in the shortest time to resume mobile services.

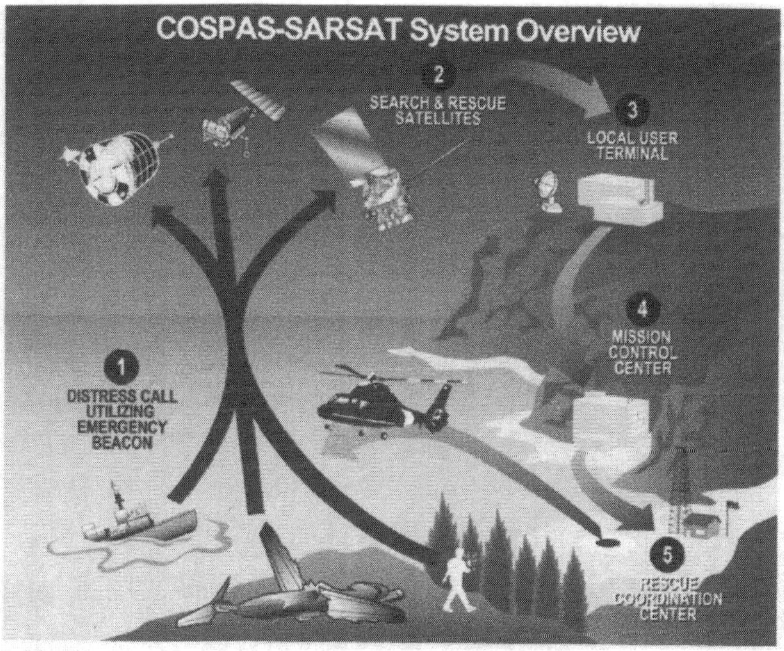

Fig.16.1 Search and rescue satellite aided tracking(SARSAT) system(NOAA-SARSAT)

Detection of microwave radiation and protection

The amount of microwave radiation at any place can be detected by an instrument which is commercially available. It can determine the extent to which we are exposed to microwave radiation at any place from mobile phones, microwave towers, computers and alarms. We can search safe places in the house where least microwave radiation is found.In order to screen out microwave radiation, we can use bed canopies made of shielded net knit material while sleeping or lying down. It is made of 100% textile material and does not require earthing. It provides maximum transparency and air for comfortable sleep.